Information Technology's Writing Survival Guide

Information Technology's Writing Survival Guide

A Technical Writing Primer for the Systems Professional

by
Patricia A. Hoyt

Information Technology's Writing Survival Guide. Copyright 2002 by Patricia A. Hoyt. All rights reserved. No part of this book may be translated, reproduced, or copied in any form, or by any mechanical or computerized means, including paper, video, audio, softbook, electronic, or information retrieval and storage systems without written permission from the publisher, except for reviewers who may quote brief passages in reviews.

ISBN 0-9679489-1-6 (printed book version)
ISBN 0-9679489-0-8 (electronic PDF file version)

LCCN 2001129313

Discounts are given for bulk purchases of five or more books. For more information visit the publisher's Web site at *www.whitefeatherpress.com*, or use the order form at the back of this book.

WhiteFeatherPress.com, Inc.
Jacksonville, Florida

About the Book

This book is an easy-to-read, comprehensive writing guide specifically for information technology professionals and for technical writers within the field of information technology. It offers methodologies for:

- Performing an audience analysis
- Working with subject matter experts
- Managing scope creep
- Planning projects
- Organizing information
- Creating outlines
- Estimating project time
- Deciding what to write

This book offers guidance on how to write:

- Executive summaries
- Processes
- Procedures
- Indexes

This book offers a review of punctuation and grammar with easy-to-understand examples *without* confusing linguistic terms. It gives numerous tips on writing style, and considerations for paper, online, or Web delivery. This book offers basic instruction on format, white space, use of color and fonts, screen design, and hyperlink design. It explains the pros and cons of using markup languages, how markup languages work, and it helps you decide which markup language would best meet your needs.

This book is the best writing guide ever written for information technology professionals, and will help you to write better and faster than ever before. Keep it on your desk for quick reference and let it make your life easier!

Disclaimer

This book is designed to provide information on technical writing and to educate readers on general methods and considerations for technical writing. It is sold with the understanding that the publisher and author are not engaged in providing advice specific to your unique technical writing needs. If legal, technical, or other services are required, the services of a competent professional should be sought.

It is not the purpose of this book to reprint or reiterate all the information that is otherwise available on the subject of technical writing, but to complement, amplify and supplement other texts. Readers are encouraged to read other available material on technical writing and to tailor all information to their specific needs.

The author and publisher performed exhaustive research to insure the accuracy and completeness of information in this book, and they assume no responsibility for errors, inaccuracies, omissions, or inconsistencies herein. Every effort has been made to make this book as complete and as accurate as possible. However, there may be typographical mistakes or content mistakes. In addition, this book contains information that was current only up to the printing date.

The author and publisher shall have neither liability nor responsibility to any person or entity with respect to any loss or damage caused, or allegedly caused, directly or indirectly by the information contained in this book, or by implementing any of the procedures or methodologies from the book. Readers are urged to seek guidance from their corporations or legal entities regarding the use of information from this book. If you do not wish to be bound by the above, you may return the book in excellent condition to the publisher for a full refund.

About the Author

Patricia A. Hoyt has served as a technical writer, system analyst, and technical consultant within the Information Technology (IT) industry for twenty years. Upon graduating with her B.A. in psychology from the State University of New York at Binghamton, she began her career with Electronic Data Systems, Inc., where she learned: IT operations in a multitiered environment, disaster recovery management, application development, technical writing, and systems life cycle management.

As a technical writer she produced paper, online reference, and Web-based manuals on: IT operations; software development; CASE methodologies; life cycle management; application testing; systems development, implementation, acquisition, retirement, and migration; disaster recovery; IT services management implementation; automated manufacturing; product recall manuals; and numerous project plans, charters, and scopes based upon the Project Management Institute's project planning methodology. She has also designed and delivered customized technical writing courses for systems professionals, and has consulted for international training corporations in developing technical writing courses for the field of IT.

As a technical consultant, Patricia later specialized in the acquisition, procurement, and implementation of very large online reference and training systems for manufacturing, IT, biotechnology, and insurance industries. For several years she specialized in making recommendations to improve the technical infrastructures of maturing development environments. Some of her areas of expertise include: technical writing, knowledge management, document management, system acquisition and implementation, and Web site design and development. Currently, Patricia designs and consults for Web site development. Originally from New York, Patricia now lives in Florida.

Acknowledgements

I have received enormous support from my family, friends, and coworkers to produce this book. I would like to give special thanks to my dear friend, editor, and proofreader, *Millie Barnes,* and to my publisher, *Ann Castle.* I am very grateful to several dear friends who offered me endless encouragement and support, especially: *Mara Tice, Bill Castro, Paula Smith,* and *Tim Jones.* I also thank each of the following people who have inspired, encouraged or supported this effort in some way: *Elaine Anwander, Gary Blake, Michael Emamdie, Debby Fairchild, Steve Jarecki, Duke Livermore, Bill Poland, Donna Royall, Mary Lou Short,* and *Mike Quintero.*

Available in Electronic Format

This book is also available in an electronic PDF file version 6.5 to be used with *Acrobat Reader*™. The electronic version is designed specifically for online use and can be accessed as a desktop application.

This powerful online guide has over 630 hyperlinks that will allow you quick access to very specific information on every topic in this printed version. You will want to keep this handy guide open as you write proposals, management reports, procedures, Web pages, executive summaries, user guides, and memos. Log on to *www.whitefeatherpress.com* and order your copy today!

Table of Contents

Chapter 1: The Importance of Planning 1
Considerations for Project Phases 3
Scope Document 9
Cultural Impacts 15
User Analyses 17
Gathering Audience Information 20

Chapter 2: Organizing Source Documents 33
Gathering Source Information 33
Procedure Definition 43
Process Definition 44
Overview Definition 45
Background Definition 45
Introduction Definition 46
Appendix Definition 47
Executive Summary Definition 47
Hierarchies of Organization 50

Chapter 3: Outlines, Tables of Contents, and Titles 55
Organizing the Outline 58
Creating the Outline 61
Writing Effective Titles 68

Chapter 4: Writing Styles 73
Word Choice 78
Use of Persons 81
Active and Passive Person 83
Gender-Inclusive Language 85

ix

Table of Contents

Use of Foreign Language 86
Using Acronyms 87
Using Jargon 90
Possession with "S" 91
Double Negatives 92

Chapter 5: Punctuation 95
Guidelines for Bulleted Lists 95
Guidelines for Numbers 99
Commas 103
Colons and Semicolons 104
Capitalization 109
Hyphens and Dashes 112
Quotations 116
Endnotes, Footnotes, and Bibliographies 117

Chapter 6: Indexing 123
How Online Indexing Impacts Systems 124
How Online Indexing Impacts Writers 127
Index Length 131
Indexing Conventions 132
Index Punctuation 136
Determining What to Index 138
Considerations for Deciding What to Index 140
Manual Indexing 143

Chapter 7: Paper Delivery 145
Paper Format 147
White Space 149
Side Labels 151

Chapter 8: Online Delivery 155
Benefits of Online Delivery 155
Different Types of Online Systems 156
Organizing Online Information 160

Table of Contents

Types of Online Information 170
Cue Cards 170
Online Procedures 172
Online Processes 175

Chapter 9: Screen Design Considerations 179
Screen Viewing Areas 182
Online Font Guidelines 188
Online Color Concepts 192

Chapter 10: Elements of Online Navigation 197
Standalone Information 199
Chunking Information 201
Hyperlink Concepts 203
Hyperlink Navigation 207

Chapter 11: Markup Language Concepts 213
Purpose of Markup Languages 214
Advantages of Using a Markup Language 218
Disadvantages of Using Markup Languages 219
SGML 219
HTML 221
XHTML 221
XML 221
WML 222

Appendices 227
Project Time Estimation Procedure 227
Common Acronyms List 234
Index 243

xi

Table of Contents

Chapter 1:

The Importance of Planning

Introduction

This chapter explains the importance of having a project plan for your documentation project. It also explains how proper planning can eliminate rewriting, and ensure that the right audience accesses the correct information.

Risks of Not Planning

When you're faced with ambitious writing deadlines, it is often difficult to resist the temptation of jumping into the writing itself without planning it. Often this behavior is encouraged in the work culture: managers hate to pass by and see you not working.

It is important to remember that without planning, you waste time and run the risk of writing content that is:

- Unorganized or incomprehensible
- For the wrong audience
- Too long or too short
- Too detailed or too limited in scope
- Inaccessible for many users

Planning a technical writing project involves much more than a detailed outline of the content. You have to be sure you're writing the correct information, and equally as important, you need to be sure readers will be able to access the information easily.

Chapter 1: The Importance of Planning

Project Plans Are Tools

As an IT professional, you already know the importance of planning a project. Project plans are simply tools to:

- Organize thoughts
- Clarify roles, responsibilities, schedules, and expectations
- Keep people focused on the same goal
- Prevent misunderstandings
- Educate and inform management about your project

Development of the plan will force you to organize your thoughts, think through detailed steps, and discover items that have the potential of being overlooked. Project plans help to ensure that nothing "slips through the cracks."

Project plans prevent misunderstandings among writers, subject-matter experts (SMEs), other teams, and management by clarifying roles, responsibilities, and time frames. Without a project plan, it is easy to forget who performs a specific task and when the task should be completed. This causes misunderstandings and confusion, and it wastes time. A good project plan helps people understand the time involved in gathering and organizing information, and makes them aware of necessary resources and key activities they might otherwise overlook.

Project Plans and Management

Project plans are valuable communication tools for you and your manager. By the nature of their positions, managers are in the habit of thinking about projects at a high, strategic level. Managers with limited experience in technical writing may not realize the time, training, resources, and dedication necessary for a technical writing project. A few managers have the erroneous idea that technical writing is no more difficult than retyping old information. Project plans provide these managers with a step-by-step view of the task at hand, and clarify the project's scope.

Chapter 1: The Importance of Planning

Get Approval

There are two reasons why management should always approve your plan before you proceed with the project. The first reason is to be sure that your management wants the project badly enough to warrant the time and effort the project will take; if they don't, it is better to stop the project before you've wasted too much time. Second, obtaining approval educates management on the time and effort the project will take. This education will also let management know what barriers may need to be resolved or the potential risks in the project.

Common Steps in a Project Plan

Most IT professionals are already familiar with the mechanics of project planning, so there is no need to reiterate how to develop a plan. It is helpful to remember that there are basic phases in a technical writing project plan that correlate to the phases in an application development project:

- Planning
- Development
- Verification
- Implementation
- Production
- Maintenance

Considerations for Project Phases

Introduction

This section offers considerations for what should be included within each phase of a technical writing project. It is intended for IT professionals who are approaching technical writing projects for the first time. Less experienced technical writers may also find this information helpful for planning writing projects in an IT environment. Most organizations have a methodology and software for planning

Chapter 1: The Importance of Planning

projects, and you should use the tools you're most comfortable with to plan your technical writing project.

Planning Phase Considerations

In the Planning Phase of a technical writing project, you gather the project and audience requirements. You identify the business reasons for creating the document, and gain a clear understanding of the audience needs and expectations. This may involve a user analysis to be sure that the needs of the audience are met. In this phase you may want to consider how to:

- Determine the business needs and scope of the documentation.
- Identify the audience and determine their reading level and reading environment.
- Gather source documents and other information.
- Identify subject-matter experts who will supply new information.
- Determine if the documentation will be delivered on paper or online.
- Determine how long the project will take and perhaps how much it will cost.
- Determine how detailed the content needs to be.

Development Phase Considerations

In the Development Phase, you analyze, organize, and write the documentation. This phase includes deciding exactly how the content should be presented. If multiple authors are involved, be sure to specify who will write what portions and agree on a common style to be edited by one person. Ideally, the finished document should read as if one person wrote it. Often there is heavy editing during this phase to ensure that the documentation is organized for a specific audience or type of document. The writing itself is edited for grammar and readability.

Chapter 1: The Importance of Planning

For paper manuals, development includes printing and binding. Be sure to find out in advance how long the printing will take. If the documentation isn't too big, you may have to make copies yourself. If the documentation will be professionally bound into a book, you will need to allow plenty of time for professional printing.

Just as in application development, this is the longest phase of the project, where most of the work is accomplished. Be realistic and generous in your assessment of how long this phase will take. Consider vacation schedules, other projects, and risks that may jeopardize your project. The *Project Time Estimation Procedure* in the *Appendices* offers a method to estimate the length of time the project will take.

Verification Phase Considerations

In the Verification Phase, subject-matter experts verify the information in the documentation. Changes are made as needed, and the document is edited for content. With online documentation, the functioning of the online system is tested. This includes testing the hyperlinks, searches, and its general accessibility. With online web presentation, the system should be tested to ensure network integrity. With small corporate intranets, response time should be also tested. If large numbers of users will access information at once, system bandwidth and server integrity may need to be monitored.

Implementation Phase Considerations

During the Implementation Phase, the documentation is delivered to the audience. Short procedures may be written and distributed to users to instruct them on how to access the system. (This can also be done during the Development Phase.) Sometimes the computer security department needs to set user security profiles to allow new users access to the online system. Sometimes Web browsers or other software needs to be installed on the users' PCs.

Chapter 1: The Importance of Planning

After the documentation is in use, you may want a process in place to obtain audience feedback, such as an e-mail address within the online system, or a telephone number. Avoid making specific people the feedback contact points: otherwise, they will receive phone calls for support long after they've left the department. Use the department's name instead.

Maintenance Phase Considerations

The Maintenance Phase is for updating the documentation and to ensure the system is working properly. There should be a process in place to make changes and ensure technical accuracy of the content, as well as a department designated to do this. In addition, there needs to be version control to identify *what* version was current *when*.

Importance of Version Control

The importance of version control becomes evident when business decisions or technical decisions depend on a certain version of guidelines that were in effect during a specific time period. For example, in the case of application development, older versions of an application and its accompanying documentation are needed to maintain the application as long as it is in production. The content may refer to source files or change management procedures that apply only to specific versions of applications. Using the wrong version of documentation to maintain an application could cause the system to fail. One easy way to maintain version control with documentation for applications is to assign the same number to the documentation as to the application. For example, help files version 2.5 would apply to version 2.5 of the application. For other kinds of documentation simply number the versions.

Life Cycle Considerations

When an application outlives its usefulness, someone needs to make the decision to retire it. This decision is often difficult and involves

Chapter 1: The Importance of Planning

many variables. The same holds true for documentation. It is often difficult to determine in advance how long a document will be valid or useful. Responsibility needs to be assigned to someone who makes that determination. Very often, it is the audience who no longer uses a document for whatever reason, and who would know if the document is still needed.

The ideal scenario would be to have a communication process in place so the audience contacts the technical writing maintenance area when the old version is no longer needed. Generally, the writer or the technical writing department is responsible for investigating if and when a document is still in use.

Consequences of Outdated Documentation

Most IT organizations ignore the consequences of outdated documentation, but it should be considered. The consequences of this are not terrible when paper manuals are used; at worst, unused manuals cause clutter, and at best they make good doorstops. However, unused or outdated online documentation can cause confusion because users expect online information to be current. In addition, when online documentation is never purged or eliminated, it becomes impossible to manage, and over time, no one knows what is important or what is current. The consequences of this are:

- Audience frustration and wasted time
- Higher costs to store huge amounts of unused information
- Poor quality work if outdated procedures and processes are used
- Conflicting information
- Inability of the organization to use the right information at the right time in response to business needs

Communicate Project Purpose Clearly

In order to be sure that your documentation is used, it is important to understand the purpose of the documentation project and to

Chapter 1: The Importance of Planning

communicate this purpose to others in the organization. The greatest technical document will have no value if no one knows it exists. In addition, a well-written technical document will not be used if it doesn't serve the right audience, or if it contains invalid or useless information. Each phase of technical writing projects has opportunities for you to communicate with subject-matter experts and management to ensure that the information is valid and necessary. Sometimes for major projects, it is helpful to announce the existence of a document to its audience, through e-mail, Web browser advertising, memos, or even posters hung on the walls. People can be very supportive of technical writing projects when they realize that the purpose of the project is to make their jobs easier.

Other Planning Considerations

To help achieve successful technical writing projects, consider the following items:

- A scope document is used to clarify ownership and the purpose of the project. (This document is discussed in the next section.)
- A user analysis takes place, or the audience is already well identified.
- The project addresses how information will be obtained, verified, and delivered.
- Subject-matter experts or management should approve the outline before writing begins.
- Necessary project resources are available.
- There is understanding of how the project impacts or interacts with other technical writing projects or application development projects.
- There is understanding of how the document impacts or links to other documents within the organization.
- Responsibility is identified for research, writing, content editing, technical editing, and maintenance of the document.
- The final document will be easily accessible and its existence will be communicated to the organization.

Chapter 1: The Importance of Planning

Project or Program?
Sometimes it is difficult to distinguish between projects and programs. Be sure you understand which you are doing before you begin planning. A project creates a specific product, such as a reference manual, and it ends at a certain date, after which resources move on to other assignments.

In contrast, a program may have a product, such as delivering training, but it never ends. For example, writing advertisements for a campaign that will continue for an indefinite amount of time is a program. Writing training material that changes with each class is a training program, not a project. Know the scope and goals of the project or program before you begin planning; otherwise, your effort will never be perceived as successful.

Scope Document

Purpose of a Scope Document
This section is intended for anyone who will be writing a large technical document. A scope document:

- Defines the objectives, goals, and scope of the project
- Identifies team members
- Identifies ownership

Depending on the size of the project, the scope document can range in size from one to ten pages. The larger and more complex the project is, the more important it is to have a scope document to steer the project in the right direction.

Scope Documents Set Expectations
At first glance, a scope document may seem like a waste of time, but many experienced technical writers use it as a tool to clarify expectations and project scope, and to identify responsibilities. The

Chapter 1: The Importance of Planning

scope document can specify what the documentation will include and just as important what it will *not* include. A scope document can serve as a record of agreements, and in some IT organizations it accompanies service level agreements, or serves as a service level agreement itself.

Reasons for a Scope Document

There are four primary reasons for cultivating the habit of writing scope documents:

1. Scope documents define how the success of the project will be measured. If the objectives of the project are never defined, everyone will have a different idea of what should be achieved, and if these expectations are not met, the project will be perceived as a failure.

2. Scope documents are good "insurance policies" when miscommunication, other teams, other projects, or events jeopardize writing projects. They remind people of their agreements.

3. Scope documents quickly communicate to new team members the history, scope, roles, and agreements of a project.

4. Scope documents serve as summaries and records of your work. If you are a technical writer by profession, scope documents serve as effective summaries of your work, which can be useful during your annual review. When you leave your position, take copies of these documents with you. They serve as a diary of the various technical writing projects you have worked on through the years, and serve as a history of your career.

Use of a Scope Document

It is helpful to have a scope document word processing template for your use. Print one out and take it with you to the first meeting when the project is discussed. Explain its purpose, and use it as a planning

Chapter 1: The Importance of Planning

guide. Strongly suggest that areas of responsibilities be identified in the first meeting. If responsibilities cannot be identified, clarify that outstanding issues need to be resolved shortly. Be sure and follow through on issue resolution after the meeting. **Do not start your project until the expectations are clearly defined!**

Content of a Scope Document

The scope document should contain the following information:

- Names, telephone numbers, locations and e-mail addresses of technical writers, subject-matter experts (SMEs), reviewers, managers, and vendors
- Purpose or type of the documentation, such as end user guide, customer service manual, IT policy manual, and so forth
- Objectives of the final document
- Estimated time that subject matter experts, reviewers, managers, or technical support will need to give to the project
- How old documentation will be converted (if applicable)
- How long it will take to edit and convert old documentation
- Estimated length of project time
- Names of contacts or other projects that might impact the writing project
- Reference material to be used during the writing

Project Risks and Constraints

The scope document should summarize any special issues, risks, or constraints that could impact the technical writing project. For example, the successful completion of an online documentation project could be contingent upon the network services area installing a larger server for the online software. The scope document should specify who will approve the content, usability, and acceptance of the final document and release you from the project. Without this acceptance, you may find yourself in a position of maintaining the content continually... forever.

Chapter 1: The Importance of Planning

Delivery Assumptions
- If the documentation will be a paper manual (hard copy), estimate how many copies will be needed, how printing and distribution will be handled, and by whom.
- If the documentation will be online (soft copy), assume what software, hardware, and access the writer will need, and also what training or access the users will need.
- If other writers will be contributing to the documentation, assumptions must be made about their level of contribution, and a common style must be used throughout the documentation.

Audience Assumptions
- If the audience is not known, a user analysis will need to be completed before documentation begins. (See the section later in this chapter on *User Analysis*.)
- If the users are already known, some assumptions might include number of users, job types, length of employment, their physical location, training requirements, their reading level, and their level of proficiency in English.

Scope Document: User Analysis

User analysis information: the second page of the scope document should contain information about the end users. It may include:
- Approximate number of users
- Job type or position
- Education level or reading level, and primary language
- Length of employment
- How users will access information and how the information will be made available to them
- Training requirements

The next two pages contain a sample of a scope document.

Chapter 1: The Importance of Planning

Example of a Scope Document

UNIX Development Guide Scope Document

Owners: Melissa Manning, Senior Technical Writer, IT Systems, PH: 12345
Shwana Smith, Technical Writer, IT Systems, PH: 12344

Purpose: This project will create a UNIX development guide to provide naming standards, library conventions, and messaging information for the UNIX platform. The guide is intended to be used in conjunction with the corporate GUI standards, and will reside in the UNIX online help files.

Reference Materials:
1. Old *UNIX Developers' Guide* from May 2000
2. "Lessons-learned" documentation from ABC project
3. UNIX vendor documentation
4. IT Organizational Strategy documentation
5. *The Messaging Architecture Procedure and Process Guide*
6. *XYZ Corporation GUI Standards*

Assumptions: Melissa, Jim, and Terry will decide the complete content of the guide and approve the table of contents before writing begins. No new information will be included in the documentation after August 31. Terry will edit for UNIX technical content. Jim will assign one of his team members to edit for architectural content; this team member will devote 25% of work time to this project. When completed, the Technical Communications Department will maintain the guide. New information should be forwarded to Shwana. Updates will be saved to the XYZHelp files nightly.

Constraints: Project may be delayed for 60-90 days if management decides to install the new version of messaging architecture software. This decision will be made later this week.

Estimated Project Time: 40 days: ECD Friday, September 10

Estimated SME Time: Terry will need to contribute approximately 1 hour per week. A member of Jim's architecture team will spend approximately 2 hours per week.

Chapter 1: The Importance of Planning

Example of a Scope Document

Delivery Assumptions: Melissa will use XYZHelp software to develop the documentation. When it's completed, she will import the documentation into the UNIX developers reference guide as a help file. Developers will access the help file through the UNIX interface and have the responsibility of having XYZHelp installed on their desktops. This can be accomplished by opening a help desk record. (Most developers already have it installed, and a mentor-training approach is recommended for developers unfamiliar with XYZHelp.) The software will reside on server XYZ007.

Audience Assumptions: Approximately 140 developers will access the documentation. They are proficient in English and read at the college level. They are all located in the Washington building and have access to the XYZ007 server. No user analysis will take place.

Other Issues: The old UNIX guide resides on paper. Shwana would like to assign another technical writer to scan the guide into a file for Melissa and delete outdated material, but no other writer is available. Melissa may have to convert and edit the old guide herself, which may take another week of time.

Contacts: Terry Baker, UNIX LAN Admin., PH: 33441
Jim McCrea, Messaging Arch., PH: 54312

Chapter 1: The Importance of Planning

Cultural Impacts

Use this section to help make assumptions about how your organization's culture could impact your project.

Hands-off Cultures

The culture of the organization can impact the length of time for a technical writing project. Some smaller or cutting-edge IT organizations have a "hands-off" culture. If you belong to this type of organization, you may be assigned a writing project and an estimated deadline. How you accomplish the project is up to you. There are few resources and few meetings, with very little approval or disapproval of your work.

The advantage of this hands-off culture is that if you know what you need to document and have current information available to you, you can often breeze through a project with little frustration. However, if you are unfamiliar with the information to be documented and need input from other people, you're often left alone without the guidance you need.

Structured Cultures

Other IT organizations are much more structured, especially in conservative industries such as government, defense, banking, or insurance. They may have several layers of management and many processes to consider. There is a chance that a manager who knows little about technical writing may micromanage your project. Many micromanagers also tend to change the scope of the project daily, and keep adding to the their list of expectations. With these managers you must take the initiative to communicate how your project should proceed, and thoroughly explain the challenges of the project. Plan a meeting to explain and justify your project plan thoroughly. Garner your manager's support and ask your manager to resolve issues for you. Keep communication positive and open, and most importantly, document your agreements.

Chapter 1: The Importance of Planning

Approval Processes

There may be an approval process or a formal project management plan for your project to follow. There may be quality-assurance processes, writing standards, design standards, and International Standards Organization (ISO) approvals to obtain. Often in large organizations, a team approves documentation before it is published. Approval processes are time consuming, so allow plenty of time for them in your plan.

Project Software

The larger the organization is the more software you may have to use in the planning and execution of the documentation. There may be a corporately approved project management software, a document management system, and a markup language to use. Most large organizations will have documentation online or accessed by a Web browser. You may need to learn online software, a markup language, markup conversion software, and understand the operation of various communication protocols. You may not be familiar with all the software packages you'll be expected to use. You'll need to consider the learning curve time in your planning.

Are You New to the Organization?

Each layer of organizational complexity adds time to the project. If you are new to a large, complex organization, don't be afraid to ask for guidance on the organization's processes. Other team members should be able to give you cultural guidance.

Most organizations fall somewhere between the two extremes of a hands-off culture and a highly structured one. Regardless of the size or type of organization you're in, you will need to make assumptions regarding how the culture will impact your project. Once those assumptions are made, you can use the *Project Time Estimation Procedure* in the Appendices to estimate how long your project will take.

Chapter 1: The Importance of Planning

User Analysis

Introduction

This section helps you determine when you may need a user analysis and offers suggestions for conducting one. It is intended for people who need a clear understanding of how their documentation will be used, and in what type of reading environment. The scope document sample in the previous section presented a homogeneous group of users with a consistent level of education who are already familiar with the online software and an older version of the documentation. Very often, however, this is not the case. Sometimes you won't know who your audience is or what documentation needs they have.

Analyzing your audience can be the single most important thing you do to ensure that your documentation will be used when it is completed. Therefore, it is very important that you know your audience well so that you can plan and write your documentation in a manner that will best serve their needs. User analyses can help you understand their needs and ensure that your documentation will actually be *used*.

Techniques for User Analyses

Surveys

This section discusses the benefits and disadvantages of using two common methods for conducting user analyses: **surveys** and **interviews.** Sending surveys to your target audience is an excellent way to learn about them. A primary benefit is that you can be as detailed as you like in your questions, and obtain information about all aspects of their needs. This may be the only way to determine some audience considerations, such as length of time on the job, level of comfort with a system, or the audience's attitude towards a new system. For example, if your organization is implementing an online reference

Chapter 1: The Importance of Planning

system for the first time, a well-written survey will reveal much about your audience's attitude towards the system. Surveys often reveal surprises. For example, you may have been told that everyone has been trained on a system, only to learn from surveys that the training was a year ago and that there has since been a 50 percent turnover in the audience.

Survey Advantages

Aside from surprises and intangible considerations (such as attitude or comfort level), surveys can reveal a need to change the document's scope or to add more information. For example, suppose you plan to write a manual for a group of developers explaining how to use a library of application program interfaces (APIs) with your messaging architecture. You have a homogenous audience, and a small, highly-technical scope. From the surveys you learn that many of the developers are contractors who are new to the organization and who need an overview of the development environment in order to use the APIs efficiently. This would impact your scope, and let you know developers need references to other sources to obtain vital information about the messaging architecture.

If you are new to the organization yourself, a survey can be an excellent way to find out about an audience, and can make an audience feel as if you really care about their needs. Most groups can be very cooperative if they are asked to fill out detailed surveys once a year. But don't over do it, as some organizations dislike surveys, or have too many surveys.

Survey Disadvantages

Surveys have drawbacks and are not effective in many situations. The primary disadvantage is that surveys are time consuming. They must be designed, distributed, filled out, returned, compiled, and analyzed. Even if they are distributed and returned electronically, they still take time to complete and process. In this era of rapid application development, time is often not a luxury.

Chapter 1: The Importance of Planning

One disadvantage of surveys is that only a small portion of them are ever returned. However, this might not be a problem if you know you have a large, homogenous audience. For example, if you survey 400 customer service representatives for a new user manual, and receive 50 detailed surveys back, you may have all the information you need.

Another disadvantage with surveys is that some organizations are replete with surveys about different topics. In this case, readers may consider surveys redundant and return them to you without the detail you asked for, or return far too few surveys to be useful. If you are uncertain of what the survey response will be like, you may not want to depend on a survey as the only way to give you the information you need. You can gain additional information by conducting interviews.

Interviewing Users

If the right people participate and if the interviews are prepared properly, interviews are the fastest way to gain detailed information about your audience. If you prepare detailed questions to ask in the interview, and invite the people who would know most about the audience's needs, you will find that an interview is an excellent way of gathering user information.

Interview Advantages

A major benefit of interviews is that they can be conducted quickly, and the information you receive will be both conceptual and detailed. If the right people are present, interviews often reveal nuances and inside information about the audience, such as the audience's past experiences with documentation or challenges with the organization's culture.

Interview Disadvantages

A disadvantage of interviews is that sometimes the information received is biased. You will receive only the opinions of the particular people you interview, and it is possible that they won't share the same

Chapter 1: The Importance of Planning

perspective as the bulk of the audience. However, getting the perspective of every potential user is rarely necessary and is often impossible unless the audience is very small.

Be well prepared for your interviews. Have pertinent questions prepared. You may even want to design a template to be used if you anticipate many interviews with future projects. People will also tend to be more responsive to your interviews if they see that you're well prepared.

Gathering Audience Information

Introduction

Use this section to remind yourself of the kinds of information needed to define your audience and to meet their documentation needs.

Define the Audience

The first step in analyzing your audience is defining who they are. If you are very familiar with your organization and the requested documentation is narrow in scope, you may need only to ask the right manager a few pertinent questions to define the audience.

Homogeneous Audience

For example, the manager may tell you that all of the users will be Web application developers with college degrees, who have all been employed with your organization at least three years. This will tell you that you can probably write at the college reading level and use the technical jargon relative to your organization. This is an example of a homogeneous audience.

Heterogeneous Audience

In other scenarios, you may need to create a single source of documentation for a very broad audience with different reading levels,

Chapter 1: The Importance of Planning

positions, and needs. For example, a document could be a comprehensive systems manual to be used by application developers, managers, and computer operators. The managers may have a college-level reading ability with the need to access only the overview, or information on the system process. The programmers may need highly technical, specific system information that includes processes, procedures, standards, and guidelines. They will likely have college degrees, but may have English as a second language. The computer operators may read at a lower reading level and need only specific procedures they can access quickly. Some of them may also have English as a second language.

Divide Document into Relevant Sections

It is easier to meet the needs of a heterogeneous audience if you divide the document into relevant groups and divide the information accordingly. Then you can target each audience in a specific section of documentation.

For example, the first chapter could be the system overview that explains the system processing and tells managers specific information they need to know about integrating the system within operations. The next few chapters could target application developers and explain in detail the technical functions and specifications. The next chapter could contain procedures specifically for day shift and night shift computer operators.

Use Audience Intention Statements

Always note at the beginning of the chapter who is your intended audience: this reference is known as an "audience intention statement." This statement immediately tells readers if they have accessed the specific information they need. Your chapters should always include an introduction that briefly states what the section is about, and who should read it. The audience intention statements need not be long: one well-worded sentence is often enough. Here are some examples:

Chapter 1: The Importance of Planning

- Application developers should reference this chapter for GUI menu standards.
- This chapter contains procedures for night-shift computer operators responsible for Initial Program Loads (IPLs) on the Amdahl host.
- Managers should reference this chapter to understand the business impact of the XYZ application processing.

Define the Chapter's Objectives

Your introduction should also tell readers the objectives of the chapter. Objectives tell readers why they should read the document and may be short-term, long-term or ongoing objectives. In the second example above, the introduction tells specific readers (computer operators) that the chapter contains procedures for an IPL, a short-term objective. In the third example above, the introduction tells a specific audience (managers) the business impact of a specific application's processing.

Reading Environment

It is important to understand the environment in which your documentation will be read. Understanding the environment will help you determine the delivery method you should employ. If it is possible, visit the reading environment. Ask yourself questions such as these:

- Will the documentation be online or on paper?
- If the documentation is on paper, will it be read while readers are using a terminal or PC? Will they be holding the manual in front of them, on their laps, or to the side of the monitor? (This should affect your layout and font size.)
- If the documentation is online, will it reside in another application session, in a drop-down box, or in a help window?
- Will the reader have many interruptions while reading?

Chapter 1: The Importance of Planning

- Will users read the documentation from a laptop or personal digital assistant? Are the users sales people in clients' offices or outside? Are they hardware engineers wedged in dark cable closets?
- Will the documentation be read from a standing or a sitting position?

At first glance, questions like these may seem trivial. However, they impact how well the information will be absorbed, and they should impact your documentation design. For example, if a computer operator is reading a paper manual from a standing position between tape mounts, the font should be large enough to be read from a couple of feet away, and the information should be vividly labeled to catch the eye quickly after an interruption.

Frequency of Use

It is also important to understand how frequently your documentation will be used. Although there are no stereotypes for any kind of audience, readers of each type of frequency often share common characteristics. Readers who have different frequencies of use include:

- Novice users
- Occasional users
- Transfer users
- Expert users

Novice Users

Novice users are those who have little or no experience with a system or skill. They are sometimes afraid of "breaking the computer" or of making mistakes, and may be too embarrassed to ask for help. They are often unable to differentiate between information that is important and information that isn't important. The sequence of documentation is very important to their understanding and needs to be well labeled. They often have trouble navigating through online documentation or large paper manuals. Novices require many cross-references,

Chapter 1: The Importance of Planning

notes, hyperlinks, and very detailed tables of contents and indexes. You may want to use a warm, easy writing style to allay their fears and make the documentation easy to read. Novices have a steep learning curve, but on the positive side, they are often curious and open to new ideas.

Occasional Users

Occasional users are those who once mastered or were trained in a skill, but who have forgotten it because they had little opportunity to use it. These are the users who have a copy of the manual... somewhere. They may remember concepts but have forgotten details. They make frequent mistakes and are often embarrassed to ask for help. They prefer using everyday language rather than technical jargon, and want notes and warnings to be highly visible.

Occasional users can be a tough audience because they are often in difficult situations. They may read a document in order to perform a task normally done by someone else. Perhaps they still have to do their own jobs as well and are under time pressure. They may be in a stressful reorganization or in an unfair situation. Perhaps they work for an insensitive supervisor who doesn't understand why it takes them two hours to do a task that someone else did in twenty minutes. Occasional users sometimes have reason to be resentful of their situation, and may also be impatient or angry with themselves for forgetting knowledge they once had.

If you are rewriting a document that has been used by occasional users, you may want to interview them. Occasional users often have useful ideas for reorganizing a document, and will often be the first readers to know what could be improved. For example, with online documentation, they often have a feel for how the hyperlinks or drop-down boxes could be improved.

Chapter 1: The Importance of Planning

Transfer Users

Transfer users are trying to transfer to a new system or situation the knowledge or skill they already have of a similar system or situation. They tend to be curious, positive, and open to suggestions. They may already have conceptual knowledge, but do not know details. They are open to asking questions and want to know new terms. They'll depend upon navigational clues, such as links, labels, indexes, and cross-references to help them find information. Sometimes transfer users are under the same stresses and expectations as occasional users, and often transfer users have never received formal training.

Expert Users

Expert users are power users who have a thorough understanding of and much experience with the system. They may ignore your documentation completely, but on occasion they may reluctantly refer to it while troubleshooting a problem. They tend to be impatient and want to accomplish a task as quickly as possible. For example, an expert will often use keystrokes rather than a mouse to access menu items.

Experts don't read; they scan. They want immediate answers. If they are using a paper manual, they will ignore the table of contents and use the index extensively, so it should be very detailed. If they are using an online system, experts will want to customize it, and they will want keystrokes available in place of mouse moves. Obviously, experts make excellent subject matter experts and content editors for your documentation.

Experts May Not Value Documentation

The hardest thing about writing for experts is that they tend to fall into one of two categories: either they don't value documentation (as *they* don't need it) or they are your best advocates for creating it and can give you excellent support.

Chapter 1: The Importance of Planning

When experts don't value documentation, it is often because they forget that not everyone is an expert, and therefore they see documentation as a waste. On rare occasions, experts can feel threatened by the accessibility of information your documentation will bring, and fear losing their status as an expert. In addition, they can inadvertently make a technical writer's position more difficult by their ability to bias management against the value of documentation.

Experts May Be Advocates for Documentation

On the other hand, experts who value documentation can be the best possible advocates of it, precisely because they're aware of certain challenges, complexities, and common pitfalls that need to be remedied by good documentation. These experts tend to be positive team players who will do their best to help you to understand and to gather information. When you encounter these advocates, garner their support, and cultivate a long-term relationship that outlasts the documentation project. Most of us can use all the friends and advocates we can get. Respect them, use them, listen to them, and count your blessings.

Other Audience Considerations

Many of the bulleted items below are helpful considerations for defining your audience:

- What are the job functions of your audience and your sub-audiences?

- What are the average education levels of your audience? This will help to determine the reading level of the audience. Some corporations and government agencies use a standard eighth-grade reading level. Most state-of-the-art word processing software and online reference software have the ability to check reading level. Find out if your documentation will be tested by specific readability software.

Chapter 1: The Importance of Planning

- Find out about your audience's average length of experience with the organization. Are they all new employees or contractors? Do they need to know preliminary information about the organization before they can use your documentation effectively? Do you need to include many references to other sources?

- Approximately how many employees will be using the documentation?

- If the online system resides on a server, will all readers have access to that server? Will they need a Web browser or other software installed? Who will be responsible for assuring they have the software?

- What is your audience's environment like? Where do they physically reside? Are they scattered in different buildings or countries? Will they be able to give undivided attention to your documentation, or will they refer to it while performing their jobs and dealing with constant interruptions? What is the atmosphere of the audience's environment? Is it relaxed and slow paced, or highly structured and micromanaged? Is it noisy?

- Are there any special audience circumstances to consider, such as foreign readers or vision-impaired readers? Will you or their management be making special considerations for them?

- Will the documentation be a paper manual, a brochure, or a quick reference guide?

- Who will be responsible for printing and distribution? What is your lead-time for printing? (Refer to *Chapter 7: Paper Delivery* for more information.)

Chapter 1: The Importance of Planning

- Will the documentation be online? Will it reside on a mainframe or client/server system? Will it be accessed locally, remotely, alone, or in tandem with other applications? Will a Web browser access it? How and when will it be backed up? Will your audience need training on the system? (Refer to *Chapter 8: Online Delivery* for more information.)

- Is there existing documentation you will be revising? How is that documentation used? Is it used occasionally or is it referenced constantly? How much detail does it have and how is it organized? Should the organization of your new documentation be changed?

Visit the Audience

Much information about your audience can be obtained by visiting them on the job. You can observe their environment, their culture and work habits, atmosphere, and work flows. Ask them questions about their needs. Answers to these types of observations can help you determine their receptivity to your document and what you can do to make your document more readable in their environment.

After you have completed your user analysis, interviewed your SMEs, and gathered your source documents, you can use the next chapter to gather your source documents, and to analyze and organize your information.

Using Subject Matter Experts

Subject matter experts (SMEs: pronounced "SMEES") are the resources who know most about the documentation topic. SMEs are the primary participants in interviews to gather information, and they should be experts on the topic and the audience. Sometimes they will be appointed to you, and sometimes you must request them. Usually, SMEs are employees or onsite consultants, but occasionally they are long-distant consultants.

Chapter 1: The Importance of Planning

SMEs should be your most valuable resource for information, and should be an integral part of the documentation project. Their primary responsibilities are to give you the correct information for your document and to edit it for content accuracy. They may also be aware of other system changes or obscure documentation that could impact your project. For example, if you are documenting how to write application program interfaces, they may be aware of some very helpful, but obscure, documentation written by a consultant a year ago.

Most SMEs truly want to help you get all the information you need, and will want to support you however they can, but in the real world, helping you with your project is generally an added responsibility for SMEs, who are very busy with their own work. Depending on the culture of your organization, it may be helpful to agree on a service level agreement with the SMEs. This will clarify roles and responsibilities, and will be good insurance for you if the SMEs seriously jeopardize your project.

Hints for Working with SMEs

Because of their own job demands, SMEs often miss meetings and are late with their input or content editing. It is helpful when dealing with SMEs to take a positive, proactive role throughout the relationship. Below is a list of helpful hints to garner the cooperation you need from SMEs:

- In the beginning of the relationship, explain your need for their support and how much you look forward to working with them. Make them understand that your documentation project is a team effort and that they need to be major team players. Be friendly, enthusiastic, and positive.

- Discuss your project with the SMEs and make them understand how important your project is, as well as its impact

Chapter 1: The Importance of Planning

on the organization. As far as possible, work with them in establishing deadlines. Let them know that you sympathize with their workload and that you really appreciate the time they will give you.

- Agree on how often you'll meet and when. Always give them an agenda a day or so before the meeting so they can prepare information to give you. After reading the agenda, the SMEs may decide to invite another person to the meeting to give expert information on a topic. Meanwhile, prepare your questions for them in advance so that the meeting will proceed quickly.

- Respect their time. If you are not prepared for a meeting, reschedule it. If they cancel or miss a meeting with you, be as understanding as possible, but also remind them of their responsibilities.

- It is helpful to have SMEs approve the project in phases, and you may want "sign-offs" (their signatures indicting approval) during key phases of the project. These sign-offs serve as quality checkpoints and help to ensure that SMEs edit what they say they will edit. After all, it is human nature for people to be more thorough when they have to sign their names to something.

Approval from SMEs

When SMEs are editing content, it is helpful to have them edit a chapter at a time and give feedback to you in increments, rather than expecting them to read and edit 100 pages at once. This will help them balance their workload better.

Importance of Sign-offs

Sign-offs help to ensure that the content of the documentation is as accurate as possible and that any errors can be found before

Chapter 1: The Importance of Planning

the documentation is delivered. Finding errors before delivery can be especially critical if your project impacts production.

Handling Poor Writing from Others

Occasionally, you will be forced to include with your documentation material that poor writers have written. Sometimes these writers will be the SMEs who wrote your source documents. Here are a few suggestions for handling these situations diplomatically:

- If their writing is okay, but lacks detail, tell them that the information is so very important that it deserves more detail in the documentation. Specifically suggest what other information should be included. Be very specific. Consider writing a detailed outline for them.

- If their writing has too much detail or goes off on tangents, tell them that their version contains important information and that sections of it will be saved or used in other sections or for future reference. Explain that for this particular documentation you'll be deleting some of the details to give the audience only the information they need at a particular point in time. Suggest that the SME's document might be useful as reference, either as an appendix or accessed by way of online hyperlinks.

- If their writing is poor, don't bruise egos by even mentioning it. Instead, tell them that their content is very valuable but that you'll be making editorial changes to their writing so that the style of the document is consistent. Then quietly clean up the writing yourself. You may be tempted to ask them to rewrite it, but a poor writer cannot improve quickly simply by being told to improve. Chances are, you will receive only a *larger amount* of poor writing than you did on the first attempt.

Chapter 1: The Importance of Planning

Thank Subject Matter Experts

Show your gratitude to the SMEs. A little thoughtfulness can have immeasurable influence in building positive relationships. Be grateful for the information the SMEs provide even if difficulties arose with them. After all, you couldn't have completed the project without their help. After their work is complete, write them a thank you note or E-mail, and be sure to send a copy to their manager.

Chapter 2:
Organizing Source Documents

Introduction

This chapter explains how to analyze and organize the information for your document. It is intended for IT professionals who are not familiar with writing projects, and for novice technical writers who need help with organization. It explains how to analyze, verify, and organize source documents, and it discusses considerations for the best way of presenting information. Read it when you are panicking about how and what to write, or when you have no idea how to get started with your project.

Gathering Source Information

Obviously, you have to gather all available source documentation before you can decide whether to include it in your new documentation. Sometimes this is more difficult than you might think. Few organizations are committed to having document repositories, document management systems, or even organized network servers on which all information on a given topic resides. If your organization uses a sophisticated document management system, enforces its use, and has a search engine that can lead you to all other documents, count yourself in the minority... and very lucky.

Generally, you'll have to have to spend time researching who might have pertinent information and where it might be stored. Most organizations have documents scattered on various servers that are

Chapter 2: Organizing Source Documents

accessed by different groups of people. There is often duplicate information, and very often much of it is outdated. Your source documents can include hundreds of files. For example, some common sources of information could be:

- The manager requesting the documentation
- Subject-matter experts (SMEs)
- Co-workers
- End users
- Librarians, or LAN administrators who are familiar with an organization's documentation storage
- Other branches of the organization
- Vendors
- In-house consultants
- Consulting organizations

After your audience analysis is completed, it is helpful to have preliminary interviews with your primary contacts to gain an understanding of the type of information that should be documented. Be sure to take notes that can guide you in analyzing the right source documents and determining what to include in your documentation.

Use a Systematic Approach

After you have gathered all the source material that is easily available, you may realize you have an enormous amount of information. It is not uncommon to have several gigabits of source documents. The task of analyzing all of it can initially seem overwhelming. Even experienced technical writers can feel overwhelmed by the magnitude of the analytical task at hand. To make matters worse, ambitious deadlines can exacerbate this feeling. Regardless of the size of the project, a systematic method can help you determine:

- What is useful?
- What is redundant?
- What needs updating?
- Where should it be stored or linked?

Sort by Title

If you have a huge amount of information to sort through, it is helpful to browse the list of titles or file names to see if you can eliminate information that is obviously not useful. If it is a large project with many source documents, look only at the titles or file names. Print a list of the titles or file names as you read them, and cross out ones you can eliminate.

Elimination Criteria

Determine basic criteria for eliminating titles. Perhaps you are interested only in files created after a certain date or on a very narrow topic. Once you know what files to use as your sources, you can begin analyzing the information.

Browse and Analyze the Information

Open each document or file and scan its table of contents if it has one. This alone can often help you to eliminate inappropriate information. Then systematically browse all of your source documents and your interview notes. Look for section titles, creation dates, and how much useful information there is. If some information is immediately useful, or glaringly lacking, note this as well.

Write yourself a short note on the information in each source file and make a "Browse List" that summarizes the information in each file. This Browse List can take a few minutes or a few days to complete, depending on how many source documents you have. As you browse, keep the writing style of the documents and other considerations in mind.

Writing Style of Source Documents

Most of the time, different people write source documents. It is helpful to spend time thinking about the benefits of each document's organization, and the degree of detail they offer. Sometimes the documents are too detailed or not detailed enough. Most times the information is either

Chapter 2: Organizing Source Documents

woefully out of date or irrelevant; if this weren't so you wouldn't be writing anything new. Sometimes you will want to use passages from source documents in your documentation if the information is still relevant. In these cases you'll want to be sure to edit the writing style to make it consistent with your own.

Considerations for Source Documents

While browsing each source document, consider these questions:
- How current is the information? Who can verify its accuracy?
- Does the information change often? If so, you may want to create hyperlinks or file links to the source document instead of inserting information that may quickly be outdated.
- Are the sources well organized? What would you change about the organization?
- Do the sources have too much detail or not enough?
- Are there special terms or jargon that should be defined or deleted?
- Are the styles of the sources consistent?
- Can the styles work well with your document without much revision?
- If the styles vary too much, can you link to those sources instead of rewriting them?
- What assumptions did the source document writers make about the audience? Are they still valid assumptions now?
- Do you understand the sources well yourself? What needs clarification?

Understand the Material

Do you understand the material well enough to write about it? A writer does not have to be a SME to write about an unfamiliar topic; a writer does need to know **how to find and organize** the right information. You will need to know enough about the subject to determine:
- What content is important?
- What content is nice to have?
- What content is irrelevant or unimportant?
- How it should be organized?
- How it will be used?

Chapter 2: Organizing Source Documents

With this information and some writing experience, you can write reasonably well about anything. The SMEs should clarify what you don't understand: that is their job. Your job is to:

- Not panic
- Obtain a comfort level with the material
- Know what to ask

Online Source Considerations

Online source documentation can pose additional questions for your project. If your source documents contain current information, you may be able to embed links from your documentation to the source. This works well if all readers have the same access; however in some organizations, they will not have the same access. You may need to eliminate duplicated information or to duplicate information.

Problems with Duplicate Information

If all readers cannot have access to the same information, you will need to duplicate the information to make it accessible for everyone. Avoid this scenario if at all possible; duplicated information is difficult to manage and a horrendous waste of system resources. If duplication of information is the only way to get your audience the information they need, then you must be sure that a *process* exists so that both copies contain consistent information and are updated at the same time. If these problems occur frequently in your organization, you may want to urge your management to purchase a document management system.

Considerations for Online Source Documents

It is easy to see how complicated using online source documentation can become. You will become aware of potential problems if you ask yourself these additional questions:

- Will any source documents be linked to my document? If so, how?
- How will changes in those documents impact mine?

Chapter 2: Organizing Source Documents

- If the sources are accessed by hyperlinks, how will the changes in those sources impact my document?
- Who will ensure that changed documents have consistent information?

Communicate Need for Document Accessibility

Without organizational online processes, a sophisticated online reference system, or a document management system in place, managing online documents in a large organization can be very difficult. It is not uncommon for an organization to have hundreds of servers, with portions of your audience not having access to some or most of them. If your documentation, the sources, and their links reside on one or more servers, communication may need to occur to ensure that all readers obtain access to the necessary servers.

This communication may be as simple as an e-mail request to network services, or an e-mail to all users requesting that they initiate the access for themselves. In other organizations the process may be more complicated; meetings may be held to justify accessibility, or to change network routing for classes of users. Try to have another person resolve these kinds of issues for you. These issues can be very time consuming and distract you from your project.

Create a Browse List

After you have sorted your sources by title and have considered how your document will interact with users in your organization, it is very helpful to create a Browse List. A Browse List summarizes the content of your source files and new information. The purpose of this list is to help you organize your document and incorporate both old and new information in the right places. If your files are all online, write a list of file names and locations as you browse; you can cut and paste these names into a word processing software or notepad. As you write your document, this will help you to insert portions of the source file (or link to it) at the correct place, without the need to search online for the file. If you browse files through a corporate intranet, you may want to create links from the Browse List to the files themselves. This may save you

38

Chapter 2: Organizing Source Documents

time later on as you write and need to refer to those files, or link your document to them. You may find it convenient to print out the list of file sources and to write manual notes about each file name to use as a Browse List. For example, you may note that one source is outdated but provides a good overview, or that one source has excellent content but is poorly written and needs revision. You may want to tape this list to your wall for immediate reference if you have many source documents. As an example, a Browse List for writing a UNIX development guide with only a few source documents might look like this:

Example of Browse List

- Old UNIX development manual, June 2000: how to guide, includes trouble shooting procedures, old HP version. Poorly written, outdated, but good sequence, paper, 50 pages.

- UNIX vendor manual, installation and administration, good for technical background but not customized for our environment, paper, 150 pages.

- Corporate network configuration diagram, current config.: shows links of UNIX to messaging arch., servers, and mainframe. Might be good for introduction, stored in t:\network\config\ diagram.ppt, 3 pages.

- Development environment guidelines, March 1999: great description of UNIX processes, explains messaging arch. well, good background on whole environment, no specific procedures, online help files, 550 KB, link to this one.

- API development guide and notes the consultant wrote last spring, current: paper, 25 pages.

- Notes from SME meetings: Joe's procedures, Terry's API issues, user analysis notes.

Chapter 2: Organizing Source Documents

Sort Browse List into Categories

When you have finished browsing your source documents and have a feel for their usefulness, you may want to sort the list of titles or file names you plan to use into general categories. For example, you could have one category for background information, and other categories for overviews, procedures, definitions, and so forth.

Analyze Missing Information

Now that your Browse List is complete, look it over. Think about what information might be missing and make a Missing Information List. This list could look like the following example:

Example of Missing Information List

- Need UNIX testing environment information, procedures, and methodology. Does Brenda in the testing lab have this?
- Need development guidelines and naming conventions.
- Need data warehouse information – warehouse didn't exist when old guide was written: what impact does it have on UNIX development now?
- What databases will be used? Will developers need database connectivity drivers?

Don't expect to find all the missing information at this point; chances are you will be adding to it as the project unfolds. Make a habit of taking your Missing Information List to meetings with your SMEs to discuss. The Browse List and the Missing Information List will help you begin analyzing the types of information you will organize.

Analyze Information Type

Introduction

This section defines different types of information commonly used in all types of technical writing. It is helpful for all writers who want to understand how to organize their particular documents. This information will help you recognize what types of information you may have before your write your outline and table of contents.

Writers Decide Information Type

In order to write about a topic, you have to analyze the information, and decide what information to put where and what to delete. Writers decide what types of information are pertinent for what documents and which chapters, and the order in which it should be presented. Before you can analyze what types of information you have, it is necessary to have a clear understanding of different information types.

People Process Different Kinds of Information

People are so adept at absorbing information quickly that we rarely stop to consider the different kinds of information we need to understand in order to do something. We are rarely aware that we are synthesizing different types of information automatically to perform common tasks. If you drive, you may have been driving for so long that you process the different types of information needed for driving without thinking about it. You can drive while listening to loud music, pondering your day's schedule, sipping coffee, and fumbling with your phone.

In effect, you have become a SME on driving, and could probably write a manual on it once you analyzed the different types of information. If you gave it some thought, you would realize that when you drive a car, you synthesize processes, procedures, rules, and principles. You may even note warnings and have references within arm's reach if needed.

For example, while driving, a **process** might be a general understanding

Chapter 2: Organizing Source Documents

of driving concepts and knowledge of how to manuever a car in heavy traffic, or an understanding of how to follow a specific route to work.

Procedures are much more detailed than processes, and contain specific actions or tasks you perform to back your car out of your garage and drive it through traffic; these actions would include shifting into reverse, turning the steering wheel, checking your views, and so forth.

I hope that you adhere to **rules** for driving, such as stopping at red traffic lights. You may follow **principles** of courteous driving, such as passing in the correct lane or not blocking intersections when you stop. You may even notice **warnings**, such as the engine light or fuel light going on as you drive. You may keep a map handy as **reference** material, in case you need it.

These types of information (processes, procedures, rules, principles, warnings, and references) are very common in all types of technical writing. Not all of them are appropriate for all documents, and other documents will need other types of information as well (such as cue cards or schematics) to impart the right knowledge at the right time.

Just-in-Time Presentation

A technical document may be well written and present the correct information, but if it doesn't present the information when it is needed, it won't be valued by your audience. A major benefit of analyzing and organizing your information before you write is to give your audience the information they need **just in time** for the task or knowledge at hand. Just-in-time presentation is common in training, user manuals, and in online reference To present information just-in-time, you need to:

- Analyze the information.
- Anticipate how the audience will need to use it and when.
- Organize the information in suitable components.
- Present the information so that it can be found quickly.

Chapter 2: Organizing Source Documents

Not all information in technical writing needs just-in-time presentation; however, if your particular document needs it, you will want to be aware of that before you organize your material. Before you can organize your outline and table of contents, you need to know who your audience is, have a general idea of the document's purpose, and have an understanding of the types of information you may wish to include. The following definitions of information types will help you determine what kind of information you have.

Procedure Definition

Procedures are the action steps a user performs to complete a task. Very often, a *procedure* is a small part of an overall *process*. Procedures are action oriented and if they are well written, the first word of a task's sentence will be a verb. Examples of task sentences are:

- **Determine the database driver needed to...**
- **Modify the code on line 32.**
- **Delete the last DD statement.**
- **Press ENTER.**

Procedures are very specific and detailed, and should mention only the physical or mental tasks the user needs to perform just in time for the next step or at that moment within the process. Procedure actions start and stop: they may repeat but they are not ongoing. Procedures should be written in active language, second person and have numbered steps. (See Chapter 4, *Writing Style and Punctuation,* for information on active language and second person.)

Usually, procedures are titled with the "ing" form of the verb. Sometimes, you may choose to use "How to..." instead, but the "ing" form of the verb is preferred. For example, some procedure titles could be:

- **How to Determine Electronic Chargebacks on the Host**
- **Determining Electronic Chargebacks on the Host**
- **Cleaning the XYZ Laser Printer**
- **Downloading Files from a Web Site**

Chapter 2: Organizing Source Documents

As you can see from the titles, procedures are for very specific actions. However, many people confuse procedures with processes, so it is important to know their differences before you organize your document.

Process Definition

Processes describe a high-level chain of events that take place over time to accomplish a goal: they may be ongoing. The intent of a written process is to give an overview of events that accomplish a goal: the intent is not to instruct users how to attain the goal. Processes contain few details and tell the user what happens in certain stages. Processes are written in third person, active voice, and often refer to many procedures. (See Chapter 4, *Writing Style and Punctuation* for information on active language and third person.)

Processes are used to give users "the big picture" of a chain of events. Procedures may be implied within the process. For example, a process for charging mainframe electronic chargeback fees, might be this:

> **Process Example**
>
> **The mainframe automatically logs usage from all cost centers, and routes the results to the Asset Management Department. Asset Management determines the chargeback fees for each cost center and routes the chargeback fees to the Accounting Department. Accounting enters the charges in each cost center's Monthly Debit Report and debits each cost center. Managers of each cost center receive the Monthly Debit Report on the first Monday of each month.**

You can see how the process explains the big picture of how electronic chargebacks are charged to each cost center. It does not instruct how to perform tasks in each phase of the process.

Processes are useful in technical writing within IT to give managers, end users, and business users an understanding of the different stages in

Chapter 2: Organizing Source Documents

accomplishing a goal. They are very often not technical in nature, although technical processes can describe technical chains of events as well.

Systems analysts often use process flows or process maps to understand the flow of data through a system or to understand the various steps in developing an application. Many people in IT are familiar with the use of flow-charting symbols used to illustrate technical processes, with each box of the flow chart representing a smaller process or a procedure. The concepts of a process (presenting the big picture for a chain of events) are the same for technical processes, business processes, or writing processes. Flow charts and processes are often major components of document overviews.

Overview Definition

Overviews are sections near the front of a document or chapter that offer general information all readers will need to know in order to understand more specific portions of the document. Very often overviews describe how the document is organized. Overviews are not very detailed, but do describe whatever is necessary for the reader to proceed with the document. Very often they contain processes and flow charts.

For example, an overview of a UNIX developer's guide would describe high-level information about the UNIX platform, how the guide is organized, and perhaps a process describing the flow of information through the system. There is sometimes a fine line between an overview and a background, as they offer similar information, but overviews do not offer an historical account of the situation.

Background Definition

Backgrounds describe an historical account of the *causes* of whatever situation your document describes. For example, if the document justifies the purchase of a large server, a background would describe the events leading up to the point where the server is necessary.

Chapter 2: Organizing Source Documents

If the document reports the findings of a technical task force, the background would describe how the team was formed and how they worked on the project.

Backgrounds are similar to overviews in that they impart general information that all readers will need to know in order to have a thorough understanding of the document. Backgrounds differ from overviews in that they always include historical information, but they do not describe how the document is organized. Often documents do not require both an overview and a background, but there is no harm in using both if you think you need to present both types of information.

Introduction Definition

Introductions explain the purpose and objectives of the document as succinctly as possible, and often define the intended audience. Introductions are often only one or two sentences (as shown in this book), and they are always the first paragraph of the document or section. An example of a more comprehensive introduction for a longer document is:

> **Example of an Introduction**
>
> This document discusses current challenges within the IT infrastructure, and how those challenges hinder the ability of the organization to satisfy critical business requirements. This document has two objectives:
>
> - To explain how a poorly defined infrastructure hinders application development
> - To recommend tactical and strategic solutions in improving the infrastructure that could benefit the entire IT organization

Notice how the introduction states the purpose and objectives: no more, no less. A background or an overview section might follow the introduction.

Chapter 2: Organizing Source Documents

Appendix Definition

An appendix is a section at the end of a document that offers additional information that is related to (but not essential to) the main body of the document. An appendix is intended for a limited audience who may require additional information. For example, in this book an appendix offers a procedure for calculating the amount of time a technical writing project may take. Often in technical documents an appendix may contain technical schematics, charts, lists of codes, or the details of information that were referenced in the main body of the document.

Appendices or *appendixes* is the plural of *appendix,* and refers to a section of more than one appendix. Appendices may contain an appendix of error codes, an appendix of charts, an appendix of schematics, and so forth.

Executive Summary Definitions

Executive summaries are short sections in the front of a document that are offer high-level, condensed information about the entire document. Very often the content follows the same chronological order as the body of the document. Many documents require an executive summary, such as acquisition proposals, technical reports, vendor requests for information, vendor requests for proposals, position papers, and cost benefit analyses, to name a few. Executive summaries are the most common type of summary in technical writing within the field of IT. Their audience is always management, although others read executive summaries as well to gain a general understanding of the document.

Executive summaries explain the highlights of the document — no more, no less. Do not include any technical details in a summary, even if the document is very technical. Include only an overview of what was done, the end results, and, if applicable, the business results and costs.

Very often, the executive summary is the only part of the document that executives read. Write the summary with this audience in mind, and anticipate what parts of the topic would be important to them. Many executives have no understanding or interest in technology, but they are

Chapter 2: Organizing Source Documents

interested in the results of an endeavor, such as its costs, its budget impact, and its return on investment. The summary should always concentrate on the expected *business* results of the project.

Executive Summary Length

Your executive summary should be no more than 10% of the total document. For example, if your document is 100 pages, the executive summary should be approximately 10 pages long. It should contain:

- The purpose of the document
- The main objectives of the system or project
- Mention of the areas involved
- A summary of results and recommendations
- A summary of costs, budget impact, and expected return on investment
- Summary of next steps

Write Executive Summaries Last

Always write executive summaries last. Executive summaries appear at the front of documents, but by writing them last you can use an easy method to write them:

1. Finalize your document.
2. Browse through it, cutting and pasting salient points in chronological order into the executive summary.
3. Heavily edit out the detail and keep in mind the 10% rule previously mentioned.
4. Change a few paragraph endings and beginnings to give the document a nice flow, and you are done.

This method will result in an excellent executive summary and be very easy to do. If you try to write the executive summary first, you will risk writing too much detail and may write irrelevant information.

Audience Impacts Document Organization

Determine how many different audiences your documentation will address. This can be accomplished by an audience analysis, as discussed in the previous chapter, and it will tell you if your audience is homogeneous or heterogeneous. Each group of readers may have different information needs. For example, some readers may need only an overview of information, while others will need detailed, just-in-time information. Your audience analysis should impact how you organize information. If one document needs to satisfy different audiences, you may want to divide your information into relevant sections according to the different information needs.

Categorize Your Information

When you have completed a Browse List, have an understanding of your audience needs, and the goals of the writing project, you have all the necessary knowledge to determine how your information should be organized in the document.

Analyze your Browse List and add new topics of information to it. Then divide the items into different categories for procedures, processes, general knowledge, graphics and photos, overviews, backgrounds, financial information, or any other categories you have. The idea is to sort the information with some kind of logic; you don't have to spend too much time on it as you'll most likely rearrange some items in the course of writing. Next, you will decide what hierarchy of information will best suit your document.

Choose a Hierarchy for Presentation

Introduction

This section explains the various types of hierarchies into which you can sort information. Regardless of what document you write and regardless of whether or not it resides on paper or online, choosing one of the following five hierarchies will allow you to present information in a systematic way that is understandable to your specific audience. After

Chapter 2: Organizing Source Documents

you have created categories for the different types of information previously mentioned, consider the following hierarchies and choose the one that would be the most logical for your specific document.

Hierarchies of Organization

Just as people are adept at learning tasks until they are automatic, we are just as adept at sorting information into hierarchies of organization without realizing it. As much as we categorize things, there are really only a few basic organizations of hierarchies:

- The least complex information to the most complex information
- The most complex information to the least complex information
- Chronological chain of events
- Sequential order
- Distinct but unrelated categories

Least-to-Most-Complex Hierarchy

This hierarchy of information is used in newspaper reporting and in business reports. It is also called the inverted pyramid approach, in which the most general and most important information is presented first, while more detailed but less important information is presented later. Most journalists use this hierarchy because it allows deletion of less important details if the article needs to be shortened. Most business reports and many financial reports are written this way.

This hierarchy of information is also very common in textbooks and user manuals. Information with simple concepts is presented first, and understanding those concepts may be necessary before more complex information can be understood. For example, an understanding of addition and subtraction is necessary before one can balance a checkbook.

If you choose this hierarchy for your particular document, you will need to understand the gradients of complexity in the information. Consult with a SME if you don't understand what components of information are more complex.

This hierarchy is a very effective for:

- Technical reports
- Business reports
- User guides
- Overviews
- Executive summaries
- Cost justification documents
- Vendor Requests for Information (RFIs)
- Vendor Requests for Proposals (RFPs)
- Design guidelines
- Explanations of various outcomes
- General reporting

Most-to-Least-Complex Hierarchy

This hierarchy is sometimes used in artistic endeavors, such as musical arrangements, but has little value for understanding quantitative information. An example of this type of hierarchy would be teaching algebra before arithmetic is mastered. You probably won't want to use this hierarchy in any type of writing.

Chronological Hierarchy

A chronological hierarchy is used when information is arranged by time. This hierarchy is probably the easiest to use in writing, as you merely have to record a chain of events. However, it works well only in a few types of documents, such as system logs, operator logs, diaries, and documents where the **timing** of information is more important than the **complexity** of information presented. Conceivably, this hierarchy would work well in some user guides where performance of tasks was very time sensitive. For example, a user guide might have tasks arranged according to the time they should be performed:

12:30 – Perform system backups.
13:45 – Perform an Initial Program Load on the mainframe.
14:00 – Restart the network.

Chapter 2: Organizing Source Documents

As a rule, however, user guides are better if they are arranged according to the least-to-most-complex hierarchy, or sequentially.

Sequential Hierarchy

A sequential hierarchy is simply arranging information according to what information is needed first, then what information is needed second, and so forth. A sequential hierarchy could also be arranged alphabetically. A sequential hierarchy is similar in its arrangement to the chronological hierarchy, and it may also be time sensitive, but it is much more adaptable to different types of documents.

Sequential hierarchies are often used in writing processes and procedures, and are very useful for user guides. Sequential hierarchies work well in other types of documents as well, such as:

- Task force reports
- Background sections
- Application development life cycle documents
- System testing guides
- Education manuals
- Software development guides
- Hardware development or maintenance guides
- Repair manuals
- Engineering manuals
- Lessons-learned documents
- Persuasive documents to gain approval

Distinct-Categories Hierarchy

A distinct-categories hierarchy presents information according to its type or category while ignoring the relationship between the categories, the complexity or sequence of information, or the importance of information. This hierarchy is simply a collection of different types of information, like a restaurant menu. Processes, procedures, overviews, or age of information might distinguish categories. An example of this hierarchy

Chapter 2: Organizing Source Documents

would be found on a Web site where users click on different categories of information (such as sales information, technical support, or updates) while the arrangement, complexity, or relationship of the categories is unimportant.

This hierarchy works very well for online documentation, online educational courses, corporate web documentation, or World Wide Web documentation. This hierarchy allows information to be updated quickly and allows for easy maintenance of the system files. For example, with this hierarchy at a Web site, information on technical support could be changed without impacting any information on sales. In addition, the system files for the Web page could be easily maintained according to file type.

The sequential hierarchy is easy to maintain but is usually not very useful in large paper documents because details have to be indexed and cross-referenced in order for users to find anything. Another disadvantage of this hierarchy is that it is difficult to explain the relationship between categories of information, or to build upon the knowledge the information imparts. Therefore, with paper delivery this hierarchy usually works well only with menus and certain catalogs.

When you finish analyzing what information you have, what information you need to gather, and the hierarchy in which it will be presented, you'll be in a good position for developing your outline and titles.

Chapter 2: Organizing Source Documents

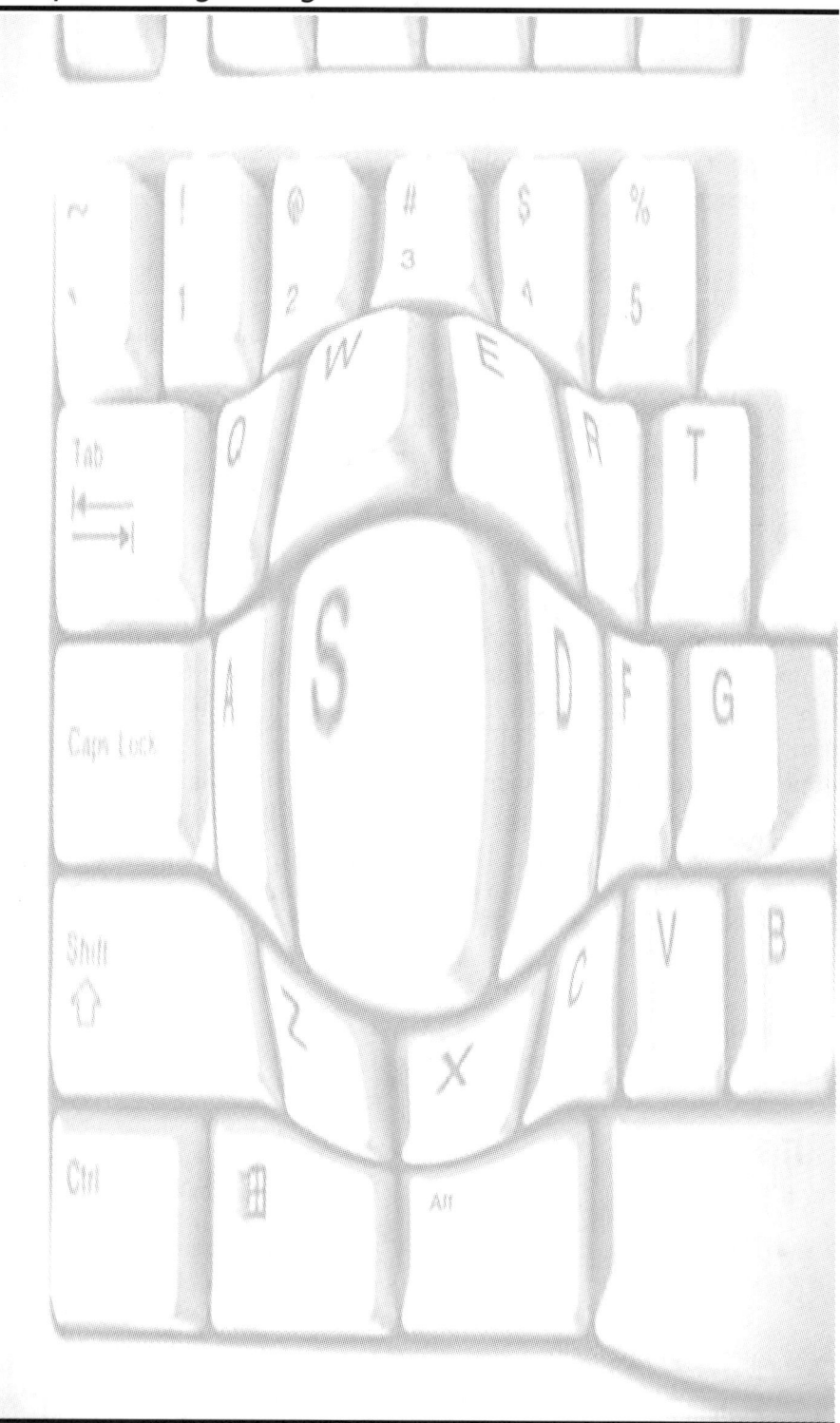

Chapter 3:
Outlines, Tables of Contents, and Titles

Introduction

This chapter discusses how to choose information for creating ourlines, tables of contents, and titles. Refer to this chapter when you need to know how to organize your information. This chapter is especially important for IT professionals who do not write large documents on a regular basis.

Reasons for Writing an Outline

This section explains how to develop an outline, how to use an outline as a tool for organizing your document and how to stay focused on your document's goal.

After you have a good understanding of what information to include in your document and what hierarchy of presentation to use, you are almost ready to begin writing. At this stage of the project it is common for one of two things to occur: either you are tempted to begin writing immediately and hope the information somehow organizes itself, or you are staring at a blank screen, overwhelmed with panic.

Outlines Help Organize Thoughts

If you are feeling overwhelmed by the size of the project and are blinded by panic at the outset of writing, an outline helps you organize your thoughts and get your arms around the task at hand. It breaks the document up into manageable components you can focus on, one at a time. More importantly, outlines help you organize the order of presentation, which usually results in a logical flow to the information.

Outlines Save Writers Time

Resist the urge to start writing immediately. Without a well-organized outline, you will spend more time reorganizing, rewriting, and moving things around than if you had taken the time to create an outline in the beginning. Outlines help writers avoid false starts by providing structure to the document. The organization of an outline helps you to include pertinent information in the right places of the document and to remain consistent. In addition, outlines often evolve into useful table of contents.

Outlines Help Writers Focus

If you are an IT professional, you are probably detailed oriented and may be an expert in your field. Without an outline, you may be tempted to write wonderful, tangential passages that are too detailed for the passage at hand, or include information in the improper sequence. For example, you may be tempted to include detailed procedures on accessing a data library within a general development process intended for a business manager. Aside from wasting time, it can be painful to delete the best writing you've ever done merely because the content's level of detail was inappropriate for the document.

Outlines Help the Flow of Information

Without an outline, you won't have a consistent focus and the flow of information in the document will suffer. Very detailed outlines can actually improve the quality of writing by allowing you to plan the flow of the document. Many experienced writers outline each paragraph to allow the information to flow smoothly from the last sentence of one paragraph, to the first sentence of the next.

Chapter 3: Outlines, Tables of Contents, and Titles

Outlines Help Writers Measure Progress

Outlines help you to judge the size of the document and your progress as you work. They give you the ability to state on a weekly status report "The XYZ document is on target. It is half done; chapter three was completed this week, and chapters four through six will be finished by the end of the month." Outlines not only give *you* a way to measure your own progress, but they give your *manager* a way to measure your progress as well.

Outlines Clarify Goals

Just as a scope document clarifies *high-level* expectations, outlines clarify the *details* of expectations. One important reason for developing an outline before you write is that the outline will clarify the document's flow and content for all team members involved. It will offer to everyone a conception of what the final document will be like before you have spent too much time writing.

Outlines Communicate Expectations

The best document will be considered a failure if it is off target or doesn't accomplish what users *perceive* it should accomplish; in writing, perception can be everything. Subject-matter experts (SMEs) are probably the best judge of what the users need, and managers can have important input. Get their opinions, solicit their advice and support, and obtain any necessary approvals of the outline before you write. This will save you wasted time on revisions, wasted time justifying why you wrote what you did, and avoid a lot of frustration. Use the outline to communicate to them what the final document will be.

Let the SMEs and managers make changes and suggestions, and give them a copy of the final outline they've approved. During the course of a project, people tend to forget what they agreed upon, and a copy of the final outline will help them remember their expectations and perceive your final document as favorably as possible.

Chapter 3: Outlines, Tables of Contents, and Titles

Organizing the Outline

Traditional Outlines

This section discusses ways to organize an outline that can later develop into a table of contents. Most likely you were taught how to create a traditional outline in school. You may remember using Roman numerals and letters that resulted in an outline something like the following example.

Example of Traditional Outline Format

Justification for a Document Management System

I. Background
A. Server Environment
    I. Information is difficult to access on 200 servers.
    II. The server information not maintained.
    III. The tactical server solutions are outlived.

B. New Growth Requires Improved Information Access.
    I. There are new business challenges with growth.
        a. There are labor shortages.
        b. There is a high turnover of employees.
    II. Recent mergers doubled the amount of information.
    III. The West coast needs access to all information.
    IV. New business requires quick access to information.
    V. New sales force requires remote access to contract information.
        a. How can we give sales force access to rapidly changing information?

II. Document Management Systems: How They Work
A. DMS access allows low-cost access by using the Web.
    I. Multiple authors can work on the same document.
    II. DMS uses XML to tag and organize information.
    III. The information can be reused in different context easily.
    IV. Web access is inexpensive.
    V. Benefits of using a DMS

Chapter 3: Outlines, Tables of Contents, and Titles

Problems with Traditional Outlines

This method of outlining probably hasn't changed much since the invention of the printing press; maybe it should. In the twenty-first century most people lose their trains of thought when deciphering a number like "XVIII." In addition, outlines written in this fashion often become projects in themselves, which is why so many people resist outlines in the first place. Few professionals have the time to perfect these types of outlines, and such a rigid numbering format can make it cumbersome to insert new ideas.

There are more modern methods available that help you outline documents more quickly than the traditional method does, and that allow you more flexibility to insert new ideas or rearrange things easily. One reason outlines can be more flexible today is that you can now employ different word processing or online reference software to organize your outline, table of contents, and index.

Basic Outline Format

Use any format you choose in which titles, topics, subtopics and notes are easily distinguishable. Many writers now ignore a detailed numbering system in favor of bullets. Bullets can be easily moved or inserted, and the indentation features of software still allow for easy leveling of detail. Leave generous space between lines, so you can insert and move items easily, as in the following example.

Chapter 3: Outlines, Tables of Contents, and Titles

> **Example of a Modern Outline Format**
>
> **Justification for a Document Management System**
>
> **Background**
> - Server environment
> - Information access from 200 servers
> - Server information not maintained
> - New Growth, Information Access
>   - New business challenges
>   - Need quick access to information
>
> **Document Management Systems: How They Work**
> - DMS access allows low-cost access by using the Web.
>   - Multiple authors can work on the same document.
>   - DMS uses XML to tag and organize information.
>   - Can reuse information
>   - Web access is inexpensive.
> - Benefits of a Document Management System
>   - Improved access
>   - Information as corporate asset

Use Your Outline as a Tool

Regardless of what method you use to create an outline, use it consistently. Remember that the outline is a tool to be used; it is not necessarily a static document, and although other people may need to approve the topics in the outline, you are the most important audience of it. The outline is for **you** so be comfortable with what you create.

Use Software as a Tool

Many online software systems contain automated methods for developing an online in a logical format. If you have such as system, consider using it, as it will help you to be more productive. You will be able to use some of the methods below whether or not you have such software available to you.

Creating the Outline

Introduction

This section walks you through the mental process of creating an outline, and offers you a procedure for determining what information is needed for your documentation. Refer to this section when you need information on how to filter out topics and analyze what is missing from your sources of information. Read this section after you have read *Chapter 2: Organizing Source Documents*. This section should be referenced after you have:

- Gathered all of your source information
- Determined a presentation hierarchy
- Determined the first draft of topics to include in your documentation

Mental Process

Experienced technical writers have a logical, analytical, mental process for creating outlines, which later become their table of contents. This process consists of:

- Creating an initial list of topics
- Verifying the topics will meet the needs of your audience
- Organizing (deleting, moving, adding) the information

For this exercise let us assume you are writing a UNIX development guide for application developers creating applications in a distributed environment. Let us assume that the first draft of topics includes:

- **UNIX development procedures**
- **Overview of UNIX development environment**
- **UNIX application program interfaces with messaging system**
- **Background**
- **Developer tool guidelines**
- **Application business modeling**

Chapter 3: Outlines, Tables of Contents, and Titles

- Executive summary
- Impact of 128-bit software tools
- Prototyping and unit testing procedures
- List of UNIX development tools
- Overview of messaging system

Verify Audience's Needs

Now, think carefully about your users. Analyze their information needs and what will be the most important to them. If your users are all experienced developers, they may not need a background. Do you really expect executives to read this document? Perhaps you don't need an executive summary either.

Adhere to Scope

Use your scope document to frame the content of your document. Just as adherence to scope improves the quality of application development, adherence to the project's scope increases the quality of the documentation, helps to ensure that it will meet the needs of the audience, and makes it easier to write.

Delete Unnecessary Items

Consider the initial topic list on the previous page. Is this the right document for developer tool guidelines? If they are already written can they be linked to, instead of duplicated? Not only is duplication a waste of effort, but keeping both copies current and synchronized is difficult to manage.

Eliminate Redundancy

Eliminate redundancy of information whenever possible. If other documents have a significant impact on your document, it is usually best to hyperlink to them if you are using an online system or reference them

Chapter 3: Outlines, Tables of Contents, and Titles

if your document will reside on paper. Delete items that are outside of your scope. Let's assume your list now looks like this:

- **UNIX development procedures**
- **Overview of UNIX development environment**
- **UNIX application program interfaces with messaging system**
- **Application business modeling**
- **Testing tools**
- **Prototyping and unit testing procedures**
- **List of UNIX development tools**
- **Overview of messaging system**

Analyze New List

Analyze the new list with your audience in mind. What is missing? In this case, wouldn't the developers also need procedures for navigation testing and integration testing? Or should all testing procedures and tools be linked to this document? Would developers need to know the process for migration to the production environment, or change management? What about building services for the UNIX environment? Ask yourself questions and continue refining your list until you think you have captured the important topics, which might be something like this:

- **UNIX development procedures**
- **UNIX application program interfaces with messaging system**
- **Building services on AIX**
- **Migration to production**
- **Overview of UNIX development environment**
- **Change management process**
- **Change management tools and utilities**
- **UNIX editors, debuggers, compilers**
- **Prototyping procedures (overview with links)**
- **Unit testing procedures (overview with links)**
- **Navigation testing procedures (overview with links)**
- **Integration testing procedures (overview with links)**
- **List of UNIX development tools**
- **Overview of messaging system**

Chapter 3: Outlines, Tables of Contents, and Titles

Organize By Usage

Now organize the list in the order in which the document will be used. If the document is intended to be a developers' guide, then you may want to organize it in the order in which it would be used through the development life cycle. Always include organizational features such as the table of contents, the index, and any appendices.

Organizing Sequentially

Developers' guides, as well as other guides, often work best when ordered sequentially, in order of how the information will be used. Notice the subtle changes in the finalized list that follows:

- Table of Contents
- Overview of UNIX development environment
- Building services on AIX
- UNIX application program interfaces with messaging system
- UNIX development procedures
- Using UNIX editors, debuggers, compilers
- Prototyping procedures (overview with links)
- Unit testing procedures (overview with links)
- Navigation testing procedures (overview with links)
- Integration testing procedures (overview with links)
- Migration to production
- Change management process
- Change management tools and utilities
- Appendices
- List of UNIX development tools
- Messaging system codes

Add Details to Outline

Add details to each item, and you will have a useful outline. You may want to include notes of where links to other sources will be. An example of the first topic with details is on the next page:

Chapter 3: Outlines, Tables of Contents, and Titles

- Overview of UNIX development environment
- Operating system overview
- Locations and administrators of servers (link to list on corporate web site)
- Business groups using UNIX platform
- Developers on west coast
- Marketing
- Network administration

Organizing by Types of Information

Organizing information sequentially works well for documents that have broad scope, or many kinds of information; even in the example above, procedures and processes were in different chapters. Documents with a narrow scope, or only a couple of types of information, may work well if they are organized by information *type*.

For example, operation manuals that contain only processes and procedures might be more accessible if they are organized by procedures in the front and processes in the back. Consider the following example of an outline for a Master Console Operator's Manual:

- Chapter 1 - Daytime Procedures
- Chapter 2 - Nighttime Procedures
- Chapter 3 - Weekend Procedures
- Chapter 4 - Month End Procedures
- Chapter 5 - Operation Policies
- Appendices
- Common Error Codes
- Year End Procedures
- Phone Numbers

Notice that the most frequently used information is in the front and that the less frequently used information is in the back. Phone numbers, which might be used frequently, are in the very back to allow quick access by being the last thing in the manual.

Chapter 3: Outlines, Tables of Contents, and Titles

Organizing Alphabetically

Sometimes you may write a document that contains mainly one type of information but is very large. An example of this could be a listing of information by state or city, or a listing of specific codes. These types of documents work well when organized alphabetically. For example:

CHAPTER 3: Corporate Web Sites Listings
- Acme.com
- Acmeinc.com
- Administration.com
- Byran.com
- Cape.com
- Dreamscometrue.com

Organizing Numerically

Occasionally, a manual may be organized numerically. This might be effective for a manual of part numbers. For example:

- **1000-1500: 32 gig hard drives**
- **1500-2000: 64 gig hard drives**
- **2000-2150: Sound cards**

Remember, the purpose of the document, as well as the way your users will want to access it, will determine the best way to organize the information.

Organizing Diverse Information

For a document that contains very diverse content, organizing it by type of information often works best. If a document were titled *Development Environment Analysis*, it might be organized as shown on the next page:

Chapter 3: Outlines, Tables of Contents, and Titles

1. Executive Summary
2. Overview of Current Development Environment
   - Corporate Strategy
   - Analysis of Current Environment
3. Internal Research Results
   - Current Development Challenges
   - Software Analysis
   - Architecture Analysis
   - Legacy Systems Analysis
   - Metrics Analysis
   - Organizational Challenges
4. External Research Results
   - Consulting Organizations
   - Benchmarking Results
5. Future Trends in Development Environments
   - Technical Trends in the Industry
   - Code Reusability
6. Recommendations
   - Tactical
   - Strategic

Completed Outline

When the outline is completed, study it carefully. Always think of how your users will access information, and see if your outline meets their needs. Remember that an outline is a guide for your use; don't be afraid to make changes. For a fresh perspective, put the outline aside and reconsider it another day.

Outline Becomes the Table of Contents

Completed outlines easily translate into tables of contents for your document, and often help you decide upon an effective title.

Chapter 3: Outlines, Tables of Contents, and Titles

Writing Effective Titles

Titles Indicate Subject and Purpose

It's obvious that titles tell readers what the main topic of the document is about. In no other area of the document will readers receive so much information about the document in as few words as in the title phrase. To be effective, titles must tell readers the subject and purpose of the document and must be brief and clear. Titles need to convey a great deal of information in as few words as possible. Some writers struggle with titles because of the enormity of titles' importance and the subtle nuance a few words can convey.

Titles Imply Nuances of Meaning

There are always at least a couple of ways to express the same thought in a title. However, even subtle implications of the document's purpose and subject can be changed by a few words. Consider the different nuances of meaning in the following examples:

1. **How to Profit from Selling Software**
2. **Software: Treasure Chest of Today's Pirate**
3. **Selling Software**

Be Specific

In order to be as informative as possible, titles should be as specific as possible. In the examples above, all three titles express the idea that software can make someone money, but they all have differences in nuance. The first title implies a guide, or instructions that the reader could use for making money from software. The second title implies that large sums of money are illegally made from selling pirated software. The third title is the shortest and the vaguest. It only implies that software can be sold, but gives no hints if it is instructional, informational, or theoretical. It is important to choose your words carefully and to consider the nuance of meaning your title conveys.

Clarify Words with Double Meanings

Choose your words carefully and clarify words with double meanings. For example, the title *Women in Labor* can have different implications to pregnant women or labor organizers, depending upon the reader's frame of reference, because of the double meaning of the word "labor." Longer, clearer titles could be *Women in the Labor Force* or *Women in Labor: Childbirth at Home.*

Decision Making Document Titles

Let your readers know from the title of the document the type of information they can expect. For example, if the document is for **decision making,** use words like:

- Proposal
- Analyzing
- Feasibility
- Justification
- Alternatives for. . .
- Options for. . .

Report Titles

If the document is a **report on a state of affairs,** use words like:

- Analysis
- Report
- Background
- Examination
- History
- Study
- State of. . .

Chapter 3: Outlines, Tables of Contents, and Titles

Tutorial Titles

If the document is a **tutorial,** use words like:

- Tutorial
- Course
- Lessons

User Guide Titles

If the document is a **user guide,** use words like:

- Manual
- Guide
- Guidelines for...
- Reference
- Users' Guide

Procedure Titles

Procedure titles can be tricky. Many people confuse processes and procedures, which are not the same thing. (Refer to *Chapter 2: Organizing Source Documents,* for clarification between procedures and processes.) If the document is for **procedures,** you have several options.

1. The preference in technical writing is to use the *ing* form of the verb, such as:

- **Troubleshooting the...**
- **Operating the...**

2. You can simply use the word "Procedure" or "Procedures," such as:

- **Procedure for Troubleshooting the...**
- **Procedures for Operations**
- **Operations Procedures**

3. Or you can use the phrase "How to..." as in:

- **How to Troubleshoot the UNIX System**
- **How to Operate the UNIX System**

Titles and Indexes

When choosing titles for procedures, consider how the titles will look in an index. For example, the phrase "How to..." could be very redundant in an index, although it may still be appropriate. Consider the number of procedures and their variety when deciding on a procedure title, but whatever phrasing you choose, be consistent with it.

Hints for Writing Effective Titles

Ask yourself these questions when deciding what a title should be:

- What key words would direct users to things they would need to know? Words like "policy," or "guidelines" are very descriptive and tell users exactly what kind of information they can expect.

- Is the document for a specific audience? If so, should this be conveyed in the title such as *Managers' Guide to....*

- Is it a title for a chapter, a topic, or a section? How much detail about the content does the title need to convey?

Answering these questions will help you write clear and effective titles that will guide a user through your manual. When you have organized your information, you can begin writing your document.

Chapter 3: Outlines, Tables of Contents, and Titles

Chapter 4:

Writing Style

Introduction

This chapter offers an overview of the types of writing styles used in modern technical writing. This chapter is not meant to be a comprehensive style guide, but offers systems professionals solutions to the most common style and grammar problems they encounter.

Writing Styles

The Purpose of Style

Writing styles are extremely important considerations in any writing project and determine how well you communicate to your selected audience. Writing styles are used to:

- Set the mood of a place or scene
- Spice up writing
- Impress readers
- Trigger emotional and intellectual responses in readers

Styles Influence Audience Response

Writing styles heavily influence both the intellectual and the emotional response your audience will have to the information. For example, when you write a proposal, your style will need to persuade your audience to support whatever it is you are proposing. Your style may need to garner their support by highlighting the benefits of your proposed ideas.

Chapter 4: Writing Style

Technical Writing Styles Emphasize Clarity

In technical writing and in business writing, style is used mainly to communicate clearly and quickly. The purpose of most technical writing is to impart specific information to a specific audience to help them perform their jobs. Simplicity and clarity should be major considerations in your style. This is especially important in today's IT environment, where many IT professionals have English as a second language. Using big words (when a simple one will do), and using Latin or foreign terms for emphasis, are now as extinct as the 286 processor.

Keep Related Words Together

Writing clearly is the single most important element in technical and business writing. If the sentence is unclear, often rearranging the words will make it clearer. Look at the following examples:

> **The Technology Charity Auction donated a hundred used PCs to local schools wrapped in protective bubble wrap.**

> **The Technology Charity Auction donated a hundred used PCs wrapped in protective bubble wrap to local schools.**

In the top example, the reader wonders if schools were wrapped in bubble wrap. In the bottom example, the meaning is made clearer by moving the phrase *wrapped in protective bubble wrap* nearer to the words to which it relates.

Limit Thoughts and Words in Sentences

Confusion often happens when too many thoughts are expressed in a single sentence. Try to limit yourself to one clear thought per sentence, and write it in twenty-five words or less; long names can be counted as one word. For example, "East Coast Network Development Department" can be counted as one word. If in doubt, shorten the sentence. It is better to have two short sentences than one long one.

Writing Difficult Sentences

If you are having trouble writing a specific sentence, you may be trying to fill it with too many thoughts. If *you* are confused, think how confused your readers will be. Try this approach:

1. Analyze exactly what you want to say.
2. Write the thoughts in *three* detailed sentences.
3. Combine two of those sentences and delete from them any unnecessary details or redundant ideas.
4. Consider leaving them as two, or even leave them as three sentences if combining them is difficult.
5. Strive for clarity of thought.

Be Specific

In technical writing, your readers learn more if you can be as specific as possible. Notice in the following examples that the bottom one offers more information that could help a technician make a more informed decision regarding network traffic:

> **The ZORO18 server has four large processors with sufficient memory, and is connected to a fast Ethernet.**

> **The ZORO18 server has four 686 processors executing at 1000 MHz. It has 512-megabytes of memory, and is connected to a 512-megabyte Ethernet.**

When deciding how specific to be, remember your audience; if your document is for executives or customers, they may care less about the technical details. For those readers, the example on the top might be better.

Use Bullets Freely

Some forms of technical information, such as a list of facts, specifications, requirements, or other items, are easier to read if a bullet precedes each item. Bullets stand out in text and help readers to remember specific information better. Often people who only scan a document will remember only the bulleted points.

For complete information on bullets and numbered lists, refer to this chapter's sections on *Guidelines for Bullet Lists* and *Guidelines for Numbered Lists*.

Delete Unnecessary Words

People frequently develop the habit of using words and phrases that do not add meaning or clarity to the thought or sentence. Editing your work carefully will help you recognize useless phrases. In the following examples, the phrases on the left can be eliminated by using the shorter phrases on the right:

in spite of the fact that	in spite of
in so much as	because
the fact that	since
she is a woman who realizes	she realizes
used for _ purposes	used for
please do not hesitate to call me	please call me
in order to	to

Use Positive Statements

Use positive and direct statements whenever possible because they are clearer in meaning. Negative statements can confuse the reader, and often seem noncommittal, as if the writer was hesitant to take a stand. The following example on the top uses unnecessary, negative words; the example on the bottom is clearer:

> **The board members were not unlike the chairman in their unrelenting drive for power.**

> **The board members were like the chairman in their relentless drive for power.**

Chapter 4: Writing Style

Use Consistent Styles

Consistent writing styles make documentation easier to write, read, and understand. Consistency gives readers subliminal expectations of what style they will be reading, which helps them absorb information faster. Consistency allows readers to concentrate on content rather than on form.

Changes in style interrupt the flow of reading. For example, in this book the word choice is simple and the reader is addressed as *you* in the second person. If I suddenly used *the readers of this book* in third person, it would interrupt the flow of your reading.

Use Consistent Spelling Conventions

There are many words in the English language that have different spellings. Some of these differences are due to dialects and traditions, such as the British spelling of "colour" and the American spelling of "color."

Other differences in spelling are due to the rapid impact of technological changes upon our culture and to the resulting changes in our language. At no other time in history has English, as well as other languages, changed so drastically and quickly as in the past ten years. Jargon (technological slang) often becomes accepted language in everyday culture. Sometimes acronyms become words in themselves. For example:

- "Television" became "T.V." and then "TV."
- "Facsimile" became "FAX" and then "fax."
- "Personal computer" became "PC."

A few words in English (according to various dictionaries) are correctly spelled two different ways. Some of these include "cancelled" and "canceled" and "zeros" and "zeroes." Whichever you choose, be consistent with its use.

Word Choice

Choose an Appropriate Vocabulary

In today's complex technological environment people need information quicker than ever before. More IT systems work is being outsourced to non-English-speaking countries, and English is (and will be) the dominant language on the Internet. After Chinese, English is the second most common language in the world: one in four people have at least a rudimentary command of it. In spite of its popularity in the business world, English is a very difficult language to learn, understand, and read, with many idioms, strange expressions, and nuances. Therefore, it is important that you choose a vocabulary, or a reading level, that is appropriate for your audience. Knowing your audience well will help you to choose a vocabulary that is appropriate and effective.

Consider how your reading would suddenly be interrupted if the level of vocabulary suddenly metamorphosed into a verbose disquisition and protracted polemic of the expediencies and inutilities of sundry writing styles. See? Do your readers a favor and use a *consistent* level of vocabulary. This will be easy to do if you have a good understanding of your audience, and if you use a software package that allows you to check reading level. The software can flag words that go above the consistent reading level in your document.

Use Simple Words

People read technical documentation to gain specific knowledge, and usually those same people must read a great deal of technical material on a regular basis. Make it easier on your readers by using simple words that express the idea clearly. In the following examples, the same thoughts are expressed. It is easy to see which of the following sentences is clearer.

> **The final aspiration of technical writing is to impart concisely explicit propositions with limpidity and a lack of embellishment.**

Chapter 4: Writing Style

The goal of technical writing is to communicate specific ideas in a clear and simple way.

Use Consistent Word Choice

Consistent word choice adds to the clarity of a document because the same words mean the same thing every time. For example, a writer using the word *client* throughout a document instead of using *client, patron,* and *customer* in different sections, helps readers focus on the document's meaning. Otherwise, readers might wonder if *client* and *patron* indicate the same person, different people, or the same people in different situations.

In the field of information technology, there are many words used in new ways or used in different ways, depending upon the dialect and the corporate culture. For example, *infrastructure* can mean the technical hardware and its software upon which systems reside, or it can mean the management and accompanying administrative processes that control the technology.

There are a number of words used in IT that you should use cautiously. Be sure you understand how your audience understands them, and use them accordingly. The following is a list of some words that can have double meanings, depending upon their common usage in a particular company:

- **Architecture**
- **Context management**
- **Context sensitive**
- **Document management**
- **Electronic commerce**
- **Enterprise**
- **Governance**
- **Heterogeneous data joins (that may not really exist)**
- **Infrastructure**
- **Internet**
- **Intranet**

Chapter 4: Writing Style

- Knowledge base
- Knowledge management
- Meta
- Paradigm
- Portal
- Procedure
- Process
- Threads
- Web (corporate intranet or Internet?)

Consider how these words, as well as many others, are used and understood in your organization. If you notice any inconsistency in their usage, always clarify them in your writing. For example, you might write "governance (the enforcement of standards)" to clarify how the term is specifically used in your document.

Use Consistent Nouns and Verbs

Consistent noun and verb usage help make documents clearer by allowing readers to know who or what is doing the action without having to think if a new person or thing is involved. Choose a noun and verb tense and stick with it, unless there is a specific reason to change it. In the following examples, the top one illustrates inconsistent noun and verb usage. You can see that the bottom example with consistent noun and verb usage is easier to understand.

> **When employees need a new card key they will fill out form 101 and will have their manager sign it. Then he should bring the form to Security Services (SS) to be processed. When SS is through processing the card key she calls the employee to pick it up.**

> **When a new card key is needed, fill out form 101 and have your manager sign it. Bring the form to Security Services (SS) to be processed. When SS Services is through processing the card key they will call you to pick it up.**

Chapter 4: Writing Style

The top example is inconsistent in noun and verb usage. It begins in future tense *(will)* and third person *(they),* then changes to conditional tense *(should)* and third person *(he).* It is unclear whether the employee or the manager should bring the form to Security Services. In addition, the writer is gender biased in assuming the sex of employees.

The example on the bottom is consistent in noun and verb usage and is written in the second person, which is the preferred person for procedures. (See the next section for an explanation of "second person.")

Use of Persons

This section offers a quick review of the three persons used in speaking and writing English, and describes the most important persons for technical writing. This section also explains why the second and third persons are most often used in technical writing. In all languages, persons describe who is speaking, (or writing) and have various verb (action-word) combinations that go with each person.

First Person
Writing or speaking with **"I"** or **"we"** is using the first person: using "I" is first person singular, and using "we" is first person plural. For example:

- **I recommend that we buy XYZ software.**
- **We should load the system now.**

Second Person
Writing or speaking with *you* doing the action in the sentence is using the second person. Sometimes the *you* is only implied. The second person can be both singular and plural. For examples:

- **You have the books.**
- **Press any key to continue.**
- **Call us for more information**.

Chapter 4: Writing Style

Importance of the Second Person

The second person is also the person most heard and most expected while *receiving instruction*. Since infancy we have been addressed in this person. When an adult said, "Be quiet." we understood this to be in the *you* form; as adults we know when the *you* is implicit. The implicit nature of the second person easily carries over into writing.

Benefits of Using the Second Person

Since the purpose of most technical documentation is to instruct, it is natural to write as if talking to the person receiving the instruction. In addition, using the second person saves words as the writer doesn't need to write *he and she* and *his and her* as the writer would if the third person were used.

Second person is always used when writing procedures. The following are examples of how the second person implies *you* when writing procedures:

- **Select** *File | Print*.
- **Press** *Clear* **to erase data.**

Third Person

Writing or speaking **about people or things** *(he, his, she, hers, it, they, them)* is using the third person. *He, his, she* or *her* are third person singular, and *they* or *them* are third person plural. Third person is used when writing a process or a policy. For example:

- **The cost benefit analysis indicates...**
- **The organization must take advantage of newer technology.**
- **The Information Technology Department researched...**
- **The Vice President of Systems directed the acquisition.**
- **They are using technology to reduce expenses.**
- **What are their e-mail addresses?**

Do Not Mix Persons

Be sure never to use third person plural when your subjects are singular. For example, do not write:

- **An employee should take their card key to Security.**

Instead write:

- **An employee should take his or her card key to Security.**

Better yet, avoid the annoying *his or her* gender combination and write the entire sentence in plural:

- **Employees should take their card keys to Security.**

Active and Passive Person

This section explains the difference between the active and the passive person. In technical writing and in modern creative writing, the active person is preferred, and often makes more of an impact in communication.

Passive Person Definition

In order to avoid the passive person, you have to be able to recognize it. The passive person is used when the action in the sentence is indirect; passive person usage is considered weak. There are several characteristics of passive person:

- The person or thing that is doing the action seems less important than the object of the sentence.
- The person or thing doing the action is often near the end of the sentence.

Chapter 4: Writing Style

- The person or thing doing the action is written after the word "by."

Active Person Definition

The active person is used when the action in the sentence is direct. The subject who is doing the action is stated first and is close to its verb or action phrase: active writing is considered more dynamic and stronger. Active person makes any kind of writing clearer and is a technical writing standard.

Examples of Active and Passive Person

Read the following examples to see the differences between active and passive person. The examples on the left use the passive person and contain more words than are necessary; notice how the word *by* is used or implied. The examples on the right use the active person and are more direct; it is easier to tell who is doing the action.

Passive Person Examples:	*Active Person Examples:*
Dolphins were taught by biologists in Key Largo to retrieve naval bombs.	**Biologists in Key Largo taught dolphins to retrieve naval bombs.**
Control of the computer's access is provided by the administrator.	The administrator provides control of the computer's access.
The online manuals are updated by our writers.	Our writers update the online manuals.
The sound barrier was broken by the fighter jets.	The fighter jets broke the sound barrier.
I was called by Terry yesterday.	Terry called me yesterday.

Gender-Inclusive Language

Introduction

Today, males and females work in all fields, and it is incorrect and awkward to assume that certain positions are filled by either gender. For example, many women are now managers, and many men are nurses and teachers. Assuming that every employee is of one sex is offensive to members of the opposite sex, for it implies that they are invisible or unimportant. When gender-inclusive language was first used in American culture in the 1980s, it was common to include females by using awkward phrases, such as the following examples, which you should avoid:

- **An operator should use his or her password.**
- **He or she should use his or her password.**
- **S/he should use her/his password.**

This type of gender-inclusive usage was grammatically correct, but constant use of *he or she* (or worse still, *s/he*) creates confusing and awkward sentences. Such usage stops the flow of reading, and forces readers to concentrate on form rather than content. This type of gender-inclusive usage is also difficult to write, as writers must constantly check that each pronoun includes both genders: forgetting one instance would be confusing and cause inconsistencies within the document.

Use Plural Forms for Gender-Inclusive Language

Use the plural form of noun and verb to avoid awkward gender-inclusive language as shown in the following examples:

- **Operators should use their passwords.**
- **Managers who file their reports should expect. . .**
- **Application developers should use their best judgment.**

Chapter 4: Writing Style

Occasionally, it will be impossible to use plural nouns simply to avoid gender-bias language. In these cases it is sometimes helpful to either:

- Reword the sentence and delete the pronoun.
- Use nouns in all instances and avoid pronouns.
- Use the third person *one* as the pronoun.

For examples:
- **An operator should use a password.**
- **Deleting one's password will result in. . .**
- **A manager who files a report should expect. . .**
- **An application developer should use good judgment.**

Use of Foreign Language

Introduction
This section explains why the use of foreign words and phrases is discouraged in modern technical writing.

The use of foreign words and phrases in our speech and writing has a long and complex history. The English language is relatively new and evolved over a span of centuries by the combining of different languages, especially the Latin, Greek, and Germanic languages. As a result, various foreign phrases or words crept into our language and writing.

Avoid Foreign Phrases
The current trend in technical writing is to avoid using foreign words whenever possible. This is especially important because our work force is becoming more ethnically diverse and more people use English as a second language. There are few foreign words for which the English language doesn't have a suitable equivalent.

For example, on the next page, the left column has commonly used foreign phrases that can be easily replaced by the phrases in the right column:

Common Foreign Phrases to Replace with English

Foreign Phrases	Replacement Phrases
et cetera or **etc.**	and so forth; and so on
i.e.	in other words
e.g.	that is; for example
pro bono	for free; without payment
in gratis	in gratitude; thankfully

When to Use Foreign Phrases

There are special cases when foreign phrases should be retained:

- Retain foreign terms if they are in direct quotes.
- Retain Latin terms used in the legal or medical professions.

Use your best judgment regarding your audiences' familiarity with these terms: it never hurts to explain the term in plain English.

Using Acronyms

Define Acronyms

The information technology (IT) industry is full of acronyms and abbreviations. Keep in mind that not all readers will know all the acronyms, so it is important to define them.

In the United States of America it is common to write out the full meaning of common acronyms and note them in parentheses, as in the paragraph above. If the acronym is well known, use it in the remainder of the text. With unusual or new acronyms, or in a long document, it is common to write out the full meaning of an acronym once on every page; this will save readers from having to refer back in the text to find their meaning.

When *Not* to Define Acronyms

Sometimes you will encounter acronyms that are better known as the *acronym* rather than as the *words* they represent. This is especially true with Internet terminology. Remember, clarity is the goal for technical writing, and in these cases it is best to use the acronym, as in the following examples:

- **Transmission Control Protocol/Internet Protocol (use *TCP/IP*)**
- **Uniform Resource Locator (use *URL*)**
- **Extensible Mark-up Language (use *XML*)**
- **Asynchronous Transfer Mode (use *ATM*)**

Common Problems with Acronyms

There is much confusion about writing acronyms. Several decades ago, periods were used between letters, as in *U.S.A.* However, punctuation tends to stop a reader's eye, so technical writers abandoned this practice years ago.

Slashes in Acronyms

Within IT, some acronyms have a slash; for example, *I/O* and *C/S*. A slash denotes the word *or*, so these acronyms really mean "input *or* output," "client *or* server." Do not use a slash unless the *or* is important in the meaning of the acronym. Slashes are often not recognized in spell-checking software, and can be misread by imaging or scanning software, or computers.

Rules for Acronyms

When writing acronyms:

- Do not use periods, slashes, or any other punctuation between letters, unless a slash is needed to indicate the word *or*.
- Use capital letters.

- Define unusual or new acronyms once per page.
- Define the phrase first, followed by the acronym in parentheses, for example *information technology (IT)*.

Example of Acronyms

Notice how the acronym is used in the following example:

Integrated Performance Support (IPS) supports customer service representatives in performing their jobs. IPS offers online reference and online training in one convenient system.

Plural Acronyms

Plural acronyms can be confusing. Some people use an apostrophe to denote a plural, as in *PC's;* however, this really means *belonging to the PC*. Notice how plural acronyms, possessive acronyms, and apostrophes are used in the following examples:

- **IPS supports customer service representatives (CSRs) in performing their jobs.**
- **The power surge damaged my PC's hard drive.**
- **The power surge damaged every one of the PCs' hard drives: all the PCs had to be replaced.**

Using Acronyms Online

When using acronyms online, it is helpful to define them on each screen or at least once in each topic. Otherwise, online users may not understand an acronym if they entered an online document after you defined it.

Hungarian Notation

Hungarian Notation is the use of lowercase and uppercase letters in the same word, as in *HunGarian*. Always retain Hungarian Notation if it is a part of a direct quote. Hungarian notation is used to:

- Identify specific uses of words in natural programming languages
- Identify keystroke actions as substitutes for a command. (For example "tAble" indicates that the user press the *a* key to access a table function.)

Using Jargon

Jargon Definition
Jargon is a specialized vocabulary often composed of acronyms, slang, and hybrid words. It is used to communicate specific ideas between people with a common technological background.

Jargon can also be defined as obscure or pretentious language. Many words now used in everyday language were at one time jargon. For example, "cell phones" and "PC" are now accepted in everyday conversation.

Four Types of Jargon
There are four types of jargon:

- Technical terms, such as *motherboard*
- Common terms or slang used with a new meaning, such as "blow out the programs to all the PCs on the network," "crash the system," or "hosed up my PC"
- Acronyms and abbreviations such as *LAN*
- New words such as *proactive*

Avoid Jargon
It is difficult for a writer to know the vocabulary of every user, and difficult to determine if the use of jargon would be meaningful to a particular audience. Words in jargon are often loosely defined, and those definitions may not be apparent to all readers. Jargon is a type of slang, and like slang it can vary from region to region. Jargon can be meaningless for readers who have English as a second language, or for readers who use different dialects of English. Therefore, avoid jargon whenever possible.

Chapter 4: Writing Style

If Jargon Is Used

If you must use jargon for clarifying something to a highly specific audience, use it cautiously. Be sure it is meaningful. It would be especially helpful to define jargon terms in a glossary for your readers.

Possession with "S"

Apostrophe S

In English, we use an apostrophe and the letter *s* to indicate that an item belongs to a noun or a pronoun; remember that a noun is a person, place, or thing. Notice how the apostrophe and an *s* denote possession in the following examples; especially note the examples with names that end with the letter *s* or with a double *s:*

- **The organization's president left.**
- **My book's title is wrong.**
- **Charles's software license expired.**
- **Bess's Web page is boring.**
- **My boss's desk is over there.**

Plural Possession and Apostrophe S

If the noun is plural and ends with an *s,* just add an apostrophe. For example:

- **The bosses' meeting began late.**
- **Many users' complaints are not heard.**
- **Every companies' computers failed throughout the city.**

Possessive Pronouns

To confuse matters more, there are certain words that indicate possession all on their own; these are called possessive pronouns. Never use an apostrophe with these words even though they end in *s:*

Chapter 4: Writing Style

- hers
- his
- ours
- theirs
- yours
- its

Its and It's

For such a short word, *it* causes a great deal of confusion. When showing possession with the word *it,* **do not** add an apostrophe. For example:

- **The bird hurt its wing.**
- **Its bandwidth was compromised.**
- **Juan knows its owner kept the keys.**

Use *it's* when you mean the contraction of *it is*. For example:

- **It's time to execute the application.**
- **It's the responsibility of Network Services.**

Double Negatives

Introduction

This section explains how to avoid the use of double negatives in sentences. Do you remember that in math that two negative signs make a positive? It's the same in English: two negative words cancel each other out, and the meaning becomes positive. Before you can avoid double-negative words, you have to know which words are already negative.

List of Negative Words

can't	none
couldn't	not
don't	nothing
hardly	scarcely
never	shouldn't
no	won't
nobody	wouldn't

Two negative words should never be used to refer to the same thing in a sentence. In the following pairs of examples, the incorrect sentences on the top should be replaced with the correct sentences on the bottom:

Don't use no double negatives. *(Wrong)*
Don't use any double negatives. *(Right)*

That woman doesn't like nobody. *(Wrong)*
That woman doesn't like anybody. *(Right)*

The computer hardly never seems to work. *(Wrong)*
The computer hardly ever seems to work. *(Right)*

I told him he couldn't have none. *(Wrong)*
I told him he couldn't have any. *(Right)*

She shouldn't have no more problems. *(Wrong)*
She shouldn't have any more problems. *(Right)*

He doesn't do nothing right. *(Wrong)*
He doesn't do anything right. *(Right)*

There is scarcely no more bread. *(Wrong)*
There is scarcely anymore bread. *(Right)*

Chapter 4: Writing Style

He won't do nothing for me. *(Wrong)*
He won't do anything for me. *(Right)*

Other Hints for Clear Writing

There are other things to consider if you want to write clearly:

- Read what you write, and rewrite if you must: if you outline first and stay organized, you will minimize rewriting.
- Don't overwrite: if the ending seems incomplete, use a closing statement to summarize the idea.
- Don't use humor, slang, or figures of speech. Slang and humor can be misunderstood, and it will always interrupt the flow of reading.
- When you are finished writing, put the document aside. Come back to it later and read it thoroughly. This will help you see errors and phrases that could be improved.

Review the next chapter on punctuation to see how punctuation enhances reading.

Chapter 5:

Punctuation

Introduction

Read this chapter to understand how to punctuate your documents. We rarely lie awake at night wondering why languages use punctuation. But in case you do wonder, the main purposes of punctuation are to:

- Help the reader recognize sequences of thought in a sentence.
- Distinguish the relationship of thoughts to each other.
- Understand who is thinking, speaking, or doing what to whom.

Guidelines for Bulleted Lists

Using Bullets

Readers often prefer reading bulleted or numbered lists because the bullets and numbers make the content noticeable and easy to read. Bullets and numbers stand out on a page and guide the eye to important items. Busy readers who scan documents especially value bullets. Use bullets freely. Look at the following examples and see for yourself which is easier to read and remember:

> **The network still requires installation of the routers, two gateways, the cache system, and the firewall.**
>
> **The network still requires installation of:**
> - **The routers**
> - **Two gateways**
> - **The cache system**
> - **The firewall**

Bulleted List Rules

All the items in the bulleted list should either be complete sentences with appropriate punctuation, or be phrases without any punctuation at the end of each item. All the items in the bulleted list can also be considered full sentences if they each complete the introductory sentence before the colon. In other words, never have punctuation at the end of the bulleted item unless each item in the list is comprised of a complete sentence, or completes an introductory sentence. Bullet lists should also:

- Be preceded by a colon in the body of the text.
- Begin with a capital letter.
- Agree in verb tense and person.

Bulleted List Options

Bulleted lists may or may not:

- Have line spaces after the colon and/or between lines.
- Follow a period or a colon.
- Be indented.
- Have sub-bullets.

Many writers prefer line spaces and indention to make the bulleted lists more noticeable; this is a very good idea if the page or text is very crowded. Whichever options you choose for your document, use them consistently.

Bulleted List Tips

Careful phrasing can make changing your document easier to read. To introduce bulleted lists, use *the following* instead of *below,* as in these examples:

(Correct) **The committee introduced the following recommendations:**
(Incorrect) **The committee introduced the recommendations below:**

To refer to previous bulleted lists, use *the previous* instead of *above,* as in these examples:

> *(Correct)* **As discussed in the previous bulleted items, . . .**
> *(Incorrect)* **As discussed in the bulleted items above, . . .**

This way, if changes to the document create a page break between the introductory sentence and the bulleted items, the word *following* or *previous* would still apply to the location of the bullets.

Also note, that lines with bullets in front of them should be referred to as *bulleted items* or *bulleted text; bullets* are the little symbol denoting the line of text. What you probably mean when referring to bulleted items are the lines of text, rather than the symbols in front of them. For example:

> *(Correct)* **As discussed in the previous bulleted items, . . .**
> *(Incorrect)* **As discussed in the bullets above, . . .**

Several examples of bulleted lists follow: they are all correct. Notice their similarities and dissimilarities, punctuation, noun and verb agreement, and the use of spacing to set the bulleted items apart from the text.

> **Bulleted List Example # 1**
>
> **The Technical Architecture Committee recommends these steps for implementing the new Internet site:**
> - **Evaluate Internet Service Providers and network security vendors.**
> - **Select a vendor.**
> - **Establish Internet standards and guidelines for users.**

Note how the verb tense in the bulleted list agrees with the main sentence.

> **Bulleted List Example # 2**
>
> The Technical Architecture Committee recommends these steps for implementing the new Internet site:
>
> - Evaluate Internet Service Providers and network security vendors.
>
> - Select a vendor.
>
> - Establish Internet standards and guidelines for users.

Note how the spacing makes each bulleted item more visible, and how each item is a complete sentence.

> **Bulleted List Example # 3**
>
> The mission of the XYZ Outsourcing Company is to provide the highest quality of technical development to our customers. This will be accomplished by:
>
> - Employing the most competently skilled employees in the market.
> - Communicating effectively with our customers.
> - Collaborating with our customers to establish strategic goals.
> - Delighting our customers with outstanding service.

Note how each bulleted item completes a sentence begun by the introductory phrase.

> **Bulleted List Example # 4**
>
> The XYZ consultants recommend the following improvements:
>
> - Information Management should aggressively implement a code reuse program that will improve utilization and enhance the current object libraries.
> - The Application Development Department should establish a corporate testing laboratory that is independent of all other application development. All applications should be tested by quality systems testers who use the:
>   - Recommended testing methodology.
>   - Corporately approved automated testing tool.

Note how subbullets are indented.

Guidelines for Numbers

Using Numbered Lists
All of the rules, tips, and guidelines for bulleted lists apply to numbered lists, except that you use numbers in place of bullets.

Use numbered lists when there is a *reason to number* each item. Often when a group will discuss a document, numbered lists facilitate discussion. For example, a person may say, "I disagree with item five and think it should say. . ." rather than having to read the item in order for the group to recognize which item is being discussed. Some writers think any bulleted list with more than seven items should be numbered, as seven is the average number of items in a list that the average person can remember. You may also want to use numbered lists when the items indicate a specific sequence of events.

Chapter 5: Punctuation

When to Spell Out Numbers
Spell out the word for the number if:

- The numbers are zero through nine, as in, *This book has five chapters.*
- A number is the first word of a sentence, as in, *Twelve copies are on the table.* (Note: many writers think one should write out the number if it is the first word in a bulleted or numbered list. I think this is optional and should depend upon how other numbers are written in the items and how many numerals there are. Most of all, your decision should depend upon readability.)
- There are two consecutive numbers, with the first number defining the second, as in, *There are twelve, 18-inch brackets in the box.*
- The number is the only number in the text, and has no hyphens, even if the number is over nine, as in *twenty.*

Compound Numbers
Standard technical writing conventions indicate that a compound number should be written, as in *twenty-five.* However, I find that these numbers often become lost in the text, and I prefer to write any number over 10 as a numeral, as in *25.* Remember to consider your audience, and to keep clarity and readability in mind.

When to Use Numerals
Write the numeral for:

- 10 or greater, as in *This book has 12 chapters.*
- A date, as in *August 13, 2002,* or *13 August 2001.*
- Before abbreviations or symbols, as in *The modulator must be set to 9cc.*
- To identify pages, figures, diagrams, or steps, as in *Refer to Figure 6.*
- When writing fractions and percents, as in *The disk is 98.2% full.*

Decimal Numbers
When using decimals:

- Use numerals for all decimals, even if less than 10, as in *6.33 meters*.
- Use decimals to show exactness, as in *4.0 centimeters*.
- If a number is less than one, include the leading zero, as in *0.25 cm*. This helps the eye to recognize a number less than one.

Ranges of Numbers
For ranges of numbers write either:

- *To* between numbers in a range, as in *10 to 20 feet*.
- A hyphen (-) for a range of numbers as in *1-9, 16-18*.

Multiple Numbers in Sentences
When there are multiple numbers in a sentence, some special rules apply:

- If one number in the sentence is 10 or greater, use all numerals, as in *We have 1 mainframe CPU and 12 servers*.
- If two numbers are consecutive, write the first number and use a numeral for the second, as in *Enter five 4-digit codes*.

Hyphens in Numbers
When you need to spell out numbers:

- Hyphenate written compound numbers through 99, as in *twenty-four* or *forty-two*.
- Hyphenate numerals attached to words of measurement, as in *8-inch paper*.
- Do not hyphenate numerals attached to abbreviated symbols of measurement, as in *9 mg*.

Letters for Large Numbers

In recent years, writers have begun using capital letters to represent thousands, millions, billions, trillions, and zillions; usually this is in writing currency, but is also used in referring to computer memory. This practice saves writing a long string of zeros. For currency amounts lower than the millions, use the numerals. Any of the options below are correct; just use them consistently. For example:

- One thousand = 1,000 = 1K = $1,000
- One million = 1,000,000 = 1M; $1M
- One billion = 1,000,000,000 = 1B; $1B
- One trillion = 1,000,000,000,000 = 1T; $1T

Computer Memory Abbreviations

There is much confusion regarding how to write abbreviations for units of computer memory. Use the abbreviations as measurements only with numerals, never in straight text without numerals. When first mentioning a unit of measurement, spell out the word if your audience is not familiar with the term, as in *kilobyte (K)*, after that use the abbreviations. Notice the different abbreviations that are acceptable and the use of hyphens in the following examples:

- Bits per second = bps; 166-bits per second; 166 bps
- Kilobit; always spell out
- Kilobits per second = Kbps; 128-kilobits per second; 128 Kbps; 100 Kbps file
- Kilobyte = 1,024-bytes; abbreviate as K; 212K
- Kilohertz = 1,000-cycles per second = 1000 kHz
- Meg; do not use
- Megabit; always spell out
- Megabyte = 1,024-kilobytes; MB; 800 MB; 1.2-megabyte disk
- Megahertz = 1 million cycles per second; mHz; 600 mHz; 600 mHz processor
- Gig; do not use
- Gigabit; always spell out
- Gigabyte = 1,024 MB; GB; 13-gigabyte; 13 GB hard disk

- Terabit; always spell out
- Terabyte = 1,024 GB (Note: This unit of measurement is still so new to technology that it is usually is spelled out; you may want to use TB for its abbreviation after spelling it out in previous text.)

Hyphenate numbers attached to words of measurement, but do not hyphenate numbers attached to abbreviations of measurement. For example, write *31-gigabytes* or *31 GB*.

Commas

Commas in a Sequence

Commas are used to separate a sequence of more than two thoughts or items listed in a sentence. A comma is preferred before the last *and* or before the last *or* in a sequence. For example:

- **The developer should be aware of the development guidelines, the application interface program library, and the latest architecture configuration.**
- **The technician replaced the system board, the central processing unit, and the hard drive.**

Don't use a comma in a sequence of only two items. For example:

- **The technician replaced the system board and the central processing unit.**
- **A customer can buy an 800 mHz hard disk or a 1000 mHz hard disk.**

Commas Separate Phrases or Clauses

Commas help to organize related thoughts in sentences. Commas separate phrases (or clauses) that have some *supporting relationship* within a sentence.

103

Chapter 5: Punctuation

For example:

- **The technical architect, two systems analysts, and a representative from Network Services, met on Tuesday to discuss problems with network traffic.**
- **Consultants used computer modeling to estimate the labor needed for the conversion effort, and to understand the ramifications of the IT labor shortage on the development environment.**

Commas in Salutations
Commas are used in salutations in business or personal letters if the writer is on a first name basis with the addressee. For example:

- **Dear Javier,**
- **Dear Karen,**

Commas and Numbers
Commas are used to make long numbers more readable and are placed every three digits, starting from the right. In Europe, a period is used. For example:

- **2,800,499**
- **$79,995**

Colons and Semicolons

Introduction
Colons and semicolons give writers a great deal of trouble. They seem to be the most misused punctuation marks in our language. This section offers clarification for using colons and semicolons correctly.

Chapter 5: Punctuation

Colon Usage Rules
Colons are used for several reasons:

- To separate a major thought and a supporting thought in a sentence
- To separate a phrase from a list that follows it
- To set off a long quotation
- To indicate a separation between: hours and minutes, minutes and seconds, or a title from its subtitle
- As a formal salutation in a business letter

Colons in Sentences
A colon means something follows that *supports* what was just written in the same sentence. An easy way to remember this is to think of the colon as two "periods" stacked on top of each other, acting as a structure that *supports* a sentence and gives it strength. Colons separate a main thought (which is expressed first) and a supporting thought that adds detail or further explains the main thought: colons *amplify* the meaning in the main thought, as in this very sentence.

Colons are often mistakenly used as commas in sentences. Whereas commas separate phrases in sentences, colons separate a *main thought* from a *supporting thought* in the same sentence. In the following examples, the main idea of the sentence is expressed first, followed by a colon and another supporting idea that adds detail or amplifies the first idea:

- **The license scan is done: it executed in three hours.**
- **Meeting attendance was good: everyone was there.**

Colons with Hours/Minutes
Colons are used to separate hours and minutes in denoting time. Colons are also used to separate minutes from seconds, milliseconds, and nanoseconds in scientific notations. For example:

- **The electromagnetic pulse at 16:41 hours lasted 00:01:22 seconds.**

105

Colons with Ratios
Colons are used to separate numbers in ratios. For example:

- The ratio of LAN Administrators to users is 1:55.
- Men outnumber women 4:1 in the field of IT.

Colons in Titles and Subtitles
Colons are used to set off subtitles from titles, or numbered paragraphs from chapters in legal documents or in the Bible. For example:

- Do-It-Yourself Brain Surgery: Neurology Made Easy
- Brown Versus Board of Education 12:22
- Nehemiah 10:3
- Application Development: The UNIX Environment

Colons in Salutations
In formal business letters, when the writer does not know the addressee, colons are used in place of commas in the salutations. (Refer also to semicolons in the next section.) For example:

- To Whom It May Concern:
- Dear Ms. Taylor:

Colons with Quotations
Colons are used to set off quotations from the rest of the text when the quotations are longer than three lines.

- The sentence introducing the quotation should end in a colon.
- The quote that follows should be indented about one half inch or one centimeter.
- Don't use quotation marks.

For example:

The Civil War brought great sadness to President Lincoln. During the Battle of Gettysburg, he wrote:

> Four score and seven years ago, our forefathers brought forth from this continent a new nation, conceived in liberty and dedicated to the proposition that all men are created equal. Now we are engaged in a great battle, which shall determine whether that nation, or any other nation so conceived and dedicated to that proposition can long endure. . .

. . . Ellipses

Notice the previous quotation ends with three periods. This is called an *ellipse* and indicates that the quotation is incomplete. Ellipses are three periods separated by singles spaces, as in ". . ." Often word processing software changes the spacing of the periods. An ellipse also indicates a gap in thought, a pause, a gap in a series, an incomplete thought, or uncertainty. For example:

- **The table cells are A1 . . . F5.**
- **We traveled through Paris, Lyons, Toulouse, Grenoble, Nice, Marseilles . . . every major city in France.**
- **John applied for that position . . . I wonder if he'll get it.**

Semicolons

Semicolons in Sentences

Semicolons are used to join two complete thoughts into a compound sentence. The second thought does not amplify or support the first thought, (as when using a colon) but is a different idea. It is easy to remember how to use semicolons if you think of them as a "super comma," somewhere between a comma (on the bottom of the semicolon) and a period (on the top of the semicolon).

Chapter 5: Punctuation

Semicolons join two complete thoughts that would *also* be correct as two *separate* sentences. For example:

- **It was a dark and stormy night; the thunder crashed over our heads and rattled our bones.**
- **It was a dark and stormy night. The thunder crashed over our heads and rattled our bones.**

Semicolons in Lists
Semicolons are also used as "super commas" *only* in complex, lengthy sentences (which are often bulleted lists) that *already contain commas* in phrases. For example:

- **The technician carried: a cellular telephone, a pager, and a laptop PC; a set of rachet wrenches; seven screw drivers; fiber optic cable mounts; and CDs for the database drivers, printer drivers, and network servers.**

Semicolons in Salutations
Semicolons are used in the salutations of business letters when the person writing the letter has had personal contact with the addressee, but is not on a first name basis with the addressee. (If the two parties are on a first-name basis, use a comma.) For example:

- **Dear Ms. Carroll;**

    I enjoyed meeting you at the convention and wanted to contact you regarding the position we discussed. . .

- **Dear Mr. Washington;**

    Thank your for you resume and letter expressing your interest in the position for Director of Network Services. . .

Capitalization

Rules for Capitalization

Always capitalize the:

- Personal pronoun *I*.
- First letter in a sentence.
- First letter of a quotation.
- First letter of proper pronouns (names of persons, places, titles, businesses, registered trade products, institutions, government bodies, and agencies).

Capitalizing People's Titles

Capitalize academic degrees and titles when used as part of people's names. Titles used without a person's name (or a company's name) are not capitalized, *except* when the title refers to the highest national, state, or church office, such as the President of the United States of America. For example:

- **Mary Willis, Vice President of Human Resources**
- **Bill Clinton, Former President of the United States of America**
- **Leroy Washington, Manager of Computer Security**
- **Colonel Melina Morgan, Ph.D.**

Notice in the following examples that the titles don't refer to specific people so they are not capitalized:

- **The presidents of area PC user groups assembled to . . .**
- **Several managers in computer security evaluated the fire wall's robust features . . .**
- **United States presidents have declared war when . . .**

Capitalize Geographic Regions

Capitalize geographical names and regions, but don't capitalize directions (north, south, east, west). For example:

- The Midwest is the grain belt of America.
- The South slowly revived after the Civil War.
- Turn south on the highway.
- Go west on Queen Victoria Road.
- Cowboys dominated the Old West.
- The Huns swept down from the north.
- The Middle East peace talks have stalled again.
- The Baltic region has known little peace.

Capitalize Special Events

Capitalize the names of historical events, periods, days, special events, countries, states, provinces, and cities. For example:

- World War II
- The French Revolution
- The Chang Dynasty
- The Baroque Period
- Christmas Day
- The Ninth Annual Chili Pepper Festival
- South Africa
- Northwest Territories
- Tokyo

Capitalize Religions, Nationalities, and Languages

Capitalize the names of nationalities, languages, and religions. For example:

- Pakistani
- Spanish
- Roman Catholic

Capitalize Names of Deities
Always capitalize the names of deities in all religions. For example:

- **God**
- **Allah**
- **Buddha**

Rules for Capitalizing Titles
Capitalize titles of any written documentation (books, chapters, magazines, articles), musical compositions, or movies. Always capitalize:

- The first word in the title (no matter what it is).
- Nouns (names of things).
- Adjectives and adverbs (words that describe things and actions).
- Verbs (action words).

For example:

- **Blue Moon Sonata for the Violin**
- **The Koran**
- **Analysis of Quantum String Theory**
- **The Monthly Network Response Report**

Other Words in Titles
Unless they are the first word in the title, don't capitalize conjunctions (*and, or,* and *but*), or prepositions (words that show the relationship of other words, such as, *for, with, to, of, from, in,* or *by*), or articles (*the, a*). For example:

- **For Whom the Bell Tolls** ("For" is the first word.)
- **A Tear for a Soldier**
- **But Angels Cry Too** ("But" is the first word.)
- **Bozo and Bubo: Two Dogs in the Country**
- **Fire from Below**

Chapter 5: Punctuation

Capitalization and Numbers

Capitalize a noun or an abbreviation before a number when it designates a written work. For example:

- **Chapter 9**
- **Book III**
- **Genesis 1:10**
- **Paragraph 28**

Hyphens and Dashes

Use of Hyphens

A hyphen is a punctuation mark represented by the shortest horizontal line on most keyboards, as in -. A hyphen is *not* a dash, which is a *double* hyphen, as in —. Hyphens are used to:

- Join two words into a compound word.
- Combine words into a new meaning.
- Attach prefixes to words.
- Break a word at the end of line so that it can continue on the next line.

Hyphenation is how many words evolve over time into compound words. For example, it was only a hundred years ago that "tomorrow" and "today" were hyphenated as "to-morrow" and "to-day."

Hyphens in Compound Words

A hyphen is used to join two or more words into one compound word. As our language constantly evolves, these compound words tend to become one word. Rapid changes in technology also create changes in our language. For example, technical writers are now replacing "on-line" with "online," although both forms are still used. When you are unsure about which version of a word to use, *remember that the tendency is to eliminate the hyphen and create a compound word.*

Chapter 5: Punctuation

Retain Hyphens in Compound Phrases
Some phrases always retain hyphens as a part of that compound word, especially if it is the name of a place, or if more than two words are joined. For example:

- **Jack-in-the-box**
- **Hurly-burly**
- **Pie-in-the-sky**
- **Perth-Amboy, New Jersey**
- **Merry-go-round**
- **State-of-the-art software**
- **Avon-on-the-Thames, England**

Hyphens and Compound Modifiers
Often, hyphens are used in multiple-word phrases right before a word they describe, or *modify*. These multiple-word phrases are called compound modifiers because they are joined together to *add meaning to the word that follows them*. For example:

- **The decision-making process was flawed.**
- **The high-level requirements were reviewed**.
- **Multiple-word phrases that describe the next word are called compound modifiers.**

Hyphens are also used to separate words and numerals. For example:

- **Sixty-words-per-minute typist**
- **Twenty-four dollars**
- **Two 5-inch screws**

Hyphens Can Change Meaning
Often hyphens are used to join multiple words when their combination changes the meaning of the words themselves. For example, the phrase *state-of-the-art* has a slightly different meaning than each of the words separately would imply.

113

As another example, the phrase *smoke-and-mirrors* commonly refers to an intention to deceive. A reader not familiar with this phrase might think it literally referred to smoke and mirrors. When using hyphens to connect and change the meaning of words used together, be certain that your audience will understand your meaning.

Common Prefixes
A prefix is a syllable attached in front of a root word. Prefixes are used to change the meaning of the word to which they are attached. Common prefixes are:

- **"Pre"** means *before:* pre-establish, precut, preplan, prepaid
- **"Anti"** means *against:* antireligious, antiwar, anti-abortion
- **"Re"** means *repeat:* renegotiate, resort (to sort again), reuse (to use again), reestablish
- **"Pro"** means *for an idea:* prochoice, protechnology, proenvironmentalism
- **"Sub"** means below: subaction, subindex

Sometimes it is diffcult to know if you should use a hyphen with a prefix or not. The rule-of-thumb is to not use a hyphen when attaching a prefix to a root word. If in doubt, consult a dictionary.

On-line or Online?
Often words evolve and hyphens are dropped in a few years, instead of a few decades. For example, most technical writers prefer the spelling of *online* to *on-line,* reasoning that the hyphen in the word interrupts the flow of reading. Therefore, for compound words with which your audience is familiar, it is recommended that you eliminate the hyphen.

Hyphens and Word Processing
Some word processing software split words at the end of a line and adds hyphens. Try to turn off this feature if you can. It is better to have an uneven margin than to split a word at the end of a line, which can be difficult to read or interrupts the flow of reading.

Common Errors with Hyphens

When you read your document for accuracy, watch for these common errors:

- If you choose to hyphenate words at the end of lines, be sure the word processing software did not add another hyphen, as in *main--frame*.
- Never hyphenate acronyms or abbreviations, as in *U-S-A*.
- Try not to isolate a portion of a word on one line of text, as in *com-puter*.
- Don't hyphenate the last word of a paragraph or page: instead, adjust the spacing of the letters so the whole word fits in the paragraph or page.

Dashes for Emphasis

A dash is twice as a long as a hyphen. Dashes are used within sentences as informal ways to emphasize a break in thought, or to indicate something that was purposely omitted. Most keyboards do not have a special character for dashes, but most word processing software will convert a double hyphen (- -) into a dash (—). Some word processing software contains the character for a dash under special fonts or character mappings.

Usually, dashes *are not used at all* in technical or business writing, as they often imply informality, emotions, or sarcasm. If you need to emphasize a point, choose your words carefully instead. If you must use dashes, use them correctly as in the following examples:

- **The silent bid ended at noon, with — being the highest bidder.**
- **Some students — especially foreign students — find the language course difficult.**
- **I'll talk to the vendor — assuming the purchase is approved — by early next week.**

Quotations

Quotations in Sentences
For some reason, people are unsure where to place quotation marks with other punctuation. Use quotations in sentences to express exactly what another person wrote or said. Precede the quotation with a comma and place the period *inside* the quotation marks. For example:

- Martin Luther King said, "I have a dream!"
- The technician said, "Go for it!"
- "Oh no! Not again!" he yelled.
- "Do I run the application at midnight?" he asked.

Single Quotations
If the quotation falls within a quotation, place *single* quotation marks around the *inside* quote. For example:

- "He yelled 'Geronimo!' and jumped out the window!" she exclaimed.

Long Quotations
Long quotations of more than three lines don't have quotation marks unless the quotation indicates dialogue. Introduce long quotations with a colon followed by a blank line in the body of the text. Indent the quotation approximately one-half inch or one centimeter and reduce the font size by one point. Do not use quotation marks to offset the quote, as that is accomplished by the indention. For example:

The XYZ Report recommends establishing a metrics program:

> Executive management should seriously consider establishing a metrics program. A metrics program would enhance the environment by offering accurate estimations of application development time, project costs, and required resources. In addition, it would add rigor to the maturing infrastructure and be required if the corporation competes for ISO 9000 certification.

Chapter 5: Punctuation

Endnotes, Footnotes, and Bibliographies

Introduction
Endnotes and footnotes are used to refer the reader to the original source, or to add a brief comment or explanation. Bibliographies are used to refer the reader to the list of your source documents.

Punctuation for endnotes, footnotes, and bibliographies has varied some over the years. Some people are old enough to remember when obscure Latin terms (such as, *ibid.*) were used to refer to a previously cited item. This practice has been abandoned for the sake of simplicity and clarity.

Endnotes
Endnotes are like footnotes except that they are grouped together at the end of the document, or sometimes at the end of each chapter. The advantage of endnotes over footnotes is that they are easier to edit when grouped, and they do not interfere with spacing on a page. The disadvantage is that they are difficult for your reader to reference without turning pages or scrolling down a screen. With today's modern word processing packages, it is preferred that you insert *footnotes* when needed and disregard endnotes altogether. Endnotes are still used when several writers merge a document, but please use them sparingly.

Endnote numbers are sequential throughout the document, regardless of whether they are grouped in chapters or at the very end of the document. They are noted at the end of the sentence by a superscript number, as in [1]. The superscript number follows other punctuation marks in the sentence, *except* if the reference is contained inside of parentheses. For example:

- **The infrastructure is immature (according to the preferred model [2]) and should be developed according to the...**

- **The network team installed sniffer software at all processing centers.** [3]

117

Chapter 5: Punctuation

Endnote References

When your endnotes are complete, create a numbered list titled, "Notes" to provide the bibliographic information. The end of this section offers examples of how to punctuate endnotes and footnotes. You may use subscript numbers (which word processing packages can create automatically) or non-subscript numbers and a period.

Footnotes on Pages

Footnotes are like endnotes except that the reference is noted on the bottom of the page instead of at the end of a chapter or document. Footnote numbers are sequential throughout the document, and are noted at the end of the sentence by a superscript number, as in [1]. Sometimes word processing software inserts a short line to set footnotes off from the body of the text. Footnotes are easier for readers to reference than endnotes, and their use is encouraged.

Endnote and Footnote Examples

Follow the punctuation below for endnotes and footnotes. List the following information:

- Number of item
- Author's last name or authors' last names
- Key words or entire title of the article, book, or reference
- Date
- Relevant page numbers

---

[1] This is an example of a footnote.

Notice that the titles of books and magazines are no longer underlined: instead book titles and magazine titles should be in italics. Article titles are in quotation marks.

1. Hoyt, *Information Technology's Writing Survival Guide*, WhiteFeatherPress.com, Inc., March 2002, 143.
2. "The Information Technology's Strategic Plan," XYZ, Inc., December 1999, 12.
3. Furr, "Fun Crafts with Cat Hair," *Cat Ball Magazine*, New York, May 10, 2001, 24.

Multiple References in Endnotes and Footnotes
If you cite multiple references from the same source, you can use a shortened version of footnotes and endnotes in subsequent footnotes and endnotes. These shortened versions would contain:

- The author's last name
- An abbreviated form of the title
- The relevant page number

If the previous examples were the first references, the following examples could be the subsequent endnotes. Notice that in the second reference the author's name was not available; in such cases simply use the title. For examples:

1. Hoyt, *Information Technology's Writing Survival Guide*, 125.
2. **Information Technology's Strategic Plan, 15.**

Internet References
Internet and intranet sites are frequently changed. For this reason, when they are referenced in endnotes or footnotes it is important to include the:

- Title of the site
- URL address
- Date it was referenced

In addition, if the site contains many pages, cite the page within quotation marks as if citing an article. For example:

- **WhiteFeatherPress.com, Inc., "Contacting the Publisher," http://www.whitefeatherpress.com, March 6, 2002.**

Web or web?
There is often confusion in how to write the word *web*. The following guidelines are commonly used by American technical writers:

- If the word refers to the Internet, the World Wide Web, or any page that is published on the Internet or World Wide Web, capitalize the *W,* as in *Web*. In addition, if the world refers to a Web-based application created for use on the Internet, (such as a Web browser), capitalize the *W*.

- If the word refers to a closed network or corporate intranet that uses Web technology but is not connected to the Internet or World Wide Web, use the lower case *w* as in *corporate web page*.

- If you are in doubt, capitalize the *W*.

Bibliographies

Purpose
Bibliographies list the sources of your references. They give credibility to what you write, and guide the reader to additional information. References in bibliographies are listed in alphabetical order. Use periods (instead of commas as you did in the endnotes and footnotes) to separate phrases, and number the references sequentially.

Use the following example as a guide to punctuating the bibliography properly. From this example, you should be able to identify easily the different types of sources used and how their punctuation differs:

Example of a Bibliography

1. Bade, Gary. Lewinsky, Karen. "An Architecture Based Upon Principles." *Nerd Quarterly*. Houston, Texas: March 2002.

2. Halliday, Paul. Teleconference on "Architecture Assessment." XYZ Consultants, Inc.. New York: January 12, 2001.

3. Hoyt,Patricia. *Integrating Multiple Vendors*. WhiteFeatherPress.com, Inc.. Jacksonville, Florida: 1999.

4. Washington, Leroy. "Distributing World-Wide Systems." http://www.dwws.com. "Architecture Theories." May 10, 2001.

Chapter 5: Punctuation

Chapter 6:

Indexing

Introduction

This chapter is intended for anyone who needs to index a long document or who wants information on:

- The different ways that software index documents.
- How indexing impacts system resources.
- How indexes should be punctuated.
- How to determine what to index.

Purpose of Indexes

Indexes were invented centuries ago. Before there were online documents and computers, indexes were the quickest way to reference information in a long document. With an index, a reader can quickly search for a subject and reference all page numbers that refer to it. Indexes still have an important function in making documentation more usable, even in this age of digital information bouncing around the planet.

Most documents you write will be accessed from some type of online reference system. Before you write it is helpful to know how online reference systems index information. Knowing this information will help you make better decisions regarding your own indexing needs.

If you are writing with a word processing application, refer to page 144 for information on how to index using a word processor. If you are writing with an online reference or web-based system, see the following section.

Chapter 6: Indexing

How Online Indexes Impact Systems

Online Reference Systems Offer Indexing

There are several major categories of online reference systems that offer automated indexing:

- Online reference (such as online manuals)
- Document management systems (which manage the contents of many online documents that have different file types, different software systems and that reside on different networks)
- Computer-based training (online courses)
- Information management systems or knowledge management systems (which manage different types of information from different systems)
- Search engines (which search for a selected term occurring in many different documents)

For simplicity's sake they will all be referred to as *online reference systems*, and most of them can generate an index for you upon request, or automatically.

Systems Create Index Files

All of these previously mentioned online reference systems have functions that tag index entries and copy them to a separate file, which is then used to generate the index. However, the different types of online reference systems can create indexes in different ways, and it is helpful to understand how your system creates an index so you can determine the impact its use will have on the entire system.

Sometimes these online reference systems use a tremendous amount of system resources (central-processing-unit time, memory, and storage) to create an index; this is especially true if the system is mainframe-based. Sometimes because of the amount of system resources used, you will prefer *not* to index a long document, especially if readers can search on topics by key words instead of referring to an index. The online reference system administrator should know exactly how the system indexes

Chapter 6: Indexing

information and impacts system resources; if not, call the vendor. It is important to know basic concepts of how online reference systems index and how the indexing impacts systems resources so you can make informed decisions regarding the indexing of online documentation.

Impact of Indexing Upon System Resources

When an online reference system automatically creates an index, the impact it will have on the system resources depends on four things:

- The volume of information in the online reference system
- How the information is compiled for the index
- How and where the index information is stored and retrieved
- Whether or not a markup language is used (Markup languages tag different types of information, for example, titles, subtitles, and paragraphs. Refer to the *Chapter 11: Markup Language Concepts,* for a more complete explanation.)

How Systems Compile Indexes

Systems can compile information for the index in several ways, and the method they use can impact traffic on a network system in different ways. Systems can perform:

- **Concordance indexing,** which tracks every single word in the document, and then alphabetizes and indexes them.
- **Key-word-in-context (KWIC) indexing,** which tracks only key words, and then alphabetizes and indexes them.
- **Limited-word indexing,** which tracks only certain words or sometimes only words of a certain length, and then alphabetizes and indexes them.
- **Markup indexing,** which tracks only information types specified by a markup language (such as tracking only words in titles and subtitles) and then alphabetizes and indexes them.
- **Title indexing,** which tracks only words in titles, and then alphabetizes and indexes them.

These types of compilations previously mentioned are ranked in order of the greatest relative impact to a system. For example, an online reference system that tracks every single word and alphabetizes the words will use much more resources than a system that tracks only words in titles.

A system that tracks every single word will often double or triple the amount of system storage needed, because the system will be creating a second copy of the text (and tagging each word) as the index is compiled. The more *work* it takes for a system to generate an index, the more system resources it will use; this may be an important consideration if your servers are overburdened.

Index Storage Impacts

How and where the index is stored also impacts system resources. In general, a system that stores the document and its index in the same place (or server) will use fewer resources than a system that stores the document in one place and stores the index in another physical location. The closer the index is to the information source it is indexing, the fewer system resources will be used to compile an index.

For example, in one scenario an online reference system can contain manuals and store the manual files and their index file on one server. In this case, accessing information from the manuals and creating an index will use few resources, as all the data is in the same place and there is little system traffic generated to create the index or move data.

In another scenario, the online reference system may be a document management system (which manages the contents of many documents in many places). With either system the indexed information may be retrieved from hundreds of different documents stored over a large network of servers. In order for the system to compile and generate an index, there must be a great deal of data traffic over the network to store, retrieve, and compile the information into one index file.

Document Management System Impacts

Usually, when a document management system is implemented, these complicated index functions are known and discussed before purchase. Most information technology (IT) organizations that purchase document management systems are quite large, they are well aware of their network traffic, and they have the tools to monitor the impact of the system upon the network. Generally, an IT organization that is mature enough to implement a document management system is able to observe and control any system impacts the software may have.

How Online Indexes Impact Writers

Introduction

The section discusses the different ways that online reference systems index and their impact upon the writer.

Concordance Indexing

Concordance indexing tracks every single word in the document. It is considered outdated and is used primarily in mainframe-based systems. This type of indexing consumes huge amounts of system resources, as the system records every word and alphabetized them. Imagine an index that listed an entry for every "the" in the text: this is exactly what concordance indexing does and why it produces a huge index. Since every word is indexed, most of the index does not contain useful information, and needs to be heavily edited before it will be user friendly.

Concordance Indexing: Writer and Reader Impact

This type of index is not easy for writers to maintain, as they must heavily edit the file. Unless the online reference system also has an excellent keyword search feature, the index will be too large and too irrelevant to be used by readers.

Chapter 6: Indexing

As previously mentioned, this type of indexing also consumes a great deal of system resources. This fact, along with the relative uselessness of the index are the reasons this type of indexing feature is generally disabled.

KWIC Indexing: Reader Impacts

Key-word-in-context indexing tracks only key words, and is used both in older mainframe-based online reference systems and in newer network-based or web-based systems. When KWIC is available in mainframe-based online reference systems, it is considered outdated and is rarely used because it consumes a great deal of system resources. The software keeps track of the important words and alphabetizes them. Usually the software has a built-in list of common words to delete from an index, but it includes all other words not on the list, or words of more than a certain length. Sometimes the owners of the online reference system can modify this list of words to delete irrelevent words.

Within the index file, each word is surrounded by the text around it. The index is displayed as thousands of lines of text with the indexed word in bold type within the line of text. This display allows readers to see the indexed term with some contextual relativity; however, scrolling through this index is very time consuming and hard on the eyes. Because this type of mainframe-based index is system-resource intensive and difficult to use, most companies that have this type of indexing software do not use this feature.

KWIC: Impact on Writers

When KWIC is used in network-based or web-based systems, it can be very useful and it requires comparatively fewer system resources than other types of systems. However, the index is not compiled automatically. In these cases, writers define what words to include in the index. This means that writers need the subject knowledge to know which words to index and take the time to index them. The indexes produced are generally very good and useful to readers because writers have a great deal of control over the content.

Limited-Word Indexing

Limited-word indexing tracks only certain words or sometimes only words of a certain length. Limited-word indexing is available in some mainframe-based online reference systems and many network-based or web-based online reference systems.

Limited-Word: Impact on Writers

With limited-word indexing, the indexing process is semi-automated, as the system tracks every word written (as in concordance indexing) to compile the index, but creating the final version is very labor intensive. Writers themselves must delete unnecessary entries, which involves reading the text for content in one file while deleting index entries in the index file. This process is very time consuming but can produce an excellent index, as writers have a great deal of control over the content.

Markup Indexing and Title Word Indexing

Markup indexing is currently the best kind of indexing technology offers. Writers tag, or markup, words to be compiled into an index. The words can be stored in a separate file and writers can choose when to generate the index. Marking up words can be time consuming, but results in a superior index, as writers can choose words and create cross references in context. Web authoring tools, document management systems, some word processing software, and many online reference systems use this kind of indexing.

Title word indexing works the same way as markup indexing, but instead of tagging words, writers tag only titles to appear in the index. This results in an index that is scanty in content and generally not very useful.

Automated Index Maintenance

The management of online reference index systems can be very tricky, and should be controlled by a person who is knowledgeable in how the particular system works, and who has the time to maintain the index properly. A poorly maintained index will frustrate readers, waste system

Chapter 6: Indexing

resources, and compromise the quality of the online reference system. In order to use an online reference index system, you should understand how the system indexes terms, especially if the system generates an index automatically.

Index Pointers

With most online reference systems and all word processing systems, indexes must be regenerated when there are changes in the content of the documentation. When an online reference system or word processing system automatically generates an index, it creates index pointers to tell itself where each referenced item is located in the document.

Index pointers are a type of system pointer: system pointers are an internal software code that the system creates to track something within itself. All pointers are invisible to both writers and readers, and cannot be changed or edited.

In many systems, a reader can use a mouse to click on indexed terms and the pointers automatically allow the system to take the reader to the referenced item. When writers delete or insert terms manually into an index without regenerating the index, no pointers are created and the system doesn't "know" the term exists.

Example of Index Pointer Problem

For example, if the phrase *rebooting the system* is automatically indexed to be on page 89 of an online manual, the system will have created a pointer that links the indexed term to page 89. If a writer now edits page 89 and moves the phrase "rebooting the system" to page 90, the system pointer is still linked to page 89, and will now either direct a reader to the wrong page or display an error message.

Index Regeneration

In order for you to synchronize the index with any changes in the document, it is necessary for you to regenerate the index after changes are made. When a system regenerates the index, all old index pointers are removed and new ones are created. In most systems, however, this also eliminates any manual changes writers made to the old version of the index, and writers will be forced to edit the index manually again.

In some newer online reference systems, document management systems, and knowledge management systems, the system "knows" when manual edits were made to the index after the index was generated. To accomplish this, the system generates a new index, while retaining the old index, and then merges them together. The system uses an internal reference system to distinguish between new and old automated entries, and any old manual entries that writers may have inserted.

It is strongly advised that one small team be responsible for maintaining an index, and that the team understand how the system compiles and regenerates the index. The team should also agree on indexing format and conventions to maintain the index consistently.

Index Length

A guideline that many technical writers use is that there should be approximately **one index page for every 20-30 pages of text** depending of course on the complexity of the document's information. Some documents contain many facts, whose entries should be listed in the index, or have information that is complex enough to warrant one index page for every 10 pages of text.

No technical document longer than 20 pages should be without an index. If the topic is important enough for one to write 20 pages about it, it is important enough to index.

Chapter 6: Indexing

Indexing Conventions

Index Format

The format of an index should be a part of every organization's style guide. If you do not have a style guide, maybe you can create one. When enough writers start using the same styles and conventions, the conventions tend to be proliferated throughout the organization. Eventually, this results in an unofficial style guide that is adopted by the organization.

The format of an index includes its:

- Columns
- Indention
- Numbering
- Font style
- Capitalization
- Punctuation

Index Columns

For both paper and online presentation, index columns should be justified on the left and ragged on the right. Use two or three columns per page for a paper index, depending on the expected length of the index phrasing. Documents that are very technical and have lengthy phrasing should have only two columns in the index.

Use only two columns per screen for an online index. Fonts are harder to read online, and it is important not to crowd screens. If the index will display in drop-down boxes, use only one column per drop-down box.

Index Indention

Indexes use indentions because they aid scanning and allow readers to see entries and subentries. An entry should begin with a capital letter and be flush with the left margin.

Chapter 6: Indexing

If the index has a subindex, enter it on the next line and indent it approximately one-half centimeter or one-quarter inch. You should be able to set tabs at indention intervals to allow for consistent indention.

If the index has several subindexes in a hierarchy, alphabetize them, and indent each entry approximately one-half centimeter or one-quarter inch on its own line, according to its hierarchy. For example:

Automobiles
   history of
   manufacturing in
      Europe
      USA

Index Page Numbers

There are two common styles for indexing page numbers. In both styles, the primary index term does not have a page number if it has subindexes. The first of the following examples is a more formal index, and uses leader dots to guide the eye to the page number, which is flush right. If you use this style, never have more than two columns on a page or a screen, and align the page numbers using tabs so that the indentions are consistent.

The second of the following examples is a more modern index, and places the page number immediately after the indexed word. The column has a ragged-right margin and there is no need to align page numbers.

First Example of Index Style

Automobiles
   history of **124**
   manufacturing in **126**
      Europe **129**

133

Chapter 6: Indexing

Second Example of Index Format

Automobiles
   history of, 124
   manufacturing in, 126
      Europe, 129

Many people think the second style is easier to read, as the eyes have less distance to move to the page number, and there is less chance of one's vision wandering to another line. This style is easier to write manually, and easier for software to generate; sometimes software doesn't space the justified margins in the formal style exactly right, even if tabs are used.

In either style, within most online reference systems, the indexed terms can be hyperlinked. (A hyperlink is a system link that allows a reader to access the term directly by clicking on it with the mouse. Hyperlinks are often denoted by a change in font color or an underscore line. The mouse pointer will often change shape when it is over a hyperlink.)

Index Font

It is customary to use the same style font in the index as in the text of the document, *but two-points smaller*. For example, if your document is 12-point Arial, use 10-point Arial in the index. Many indexing software packages do this automatically. Never choose a font smaller than 9 points for online indexes.

Index Cross Reference

It is customary to use italics within an index to indicate a cross reference. For example:

Automobiles. *See also* **cars.**

Use a bold font to indicate that the index has an illustration or photo, or to indicate the page number for the definition of the index. In the following example, a reader would expect to find the term "automobile" *defined* on page 124:

Automobiles, **124,** 126, 140

Index Capitalization

Capitalize the first letter of an index, and proper nouns. (Remember, a proper noun is the specific name of a person, place, or thing.) Do not capitalize the first letters of subindexes unless they are proper nouns. For example:

Civil rights activists
    leaders, 119
    King, Martin Luther, 121

Notice in the following example that the subindexes under "Planets" are alphabetized, and that another level of subindex, "Rings of Saturn, 13," is indented even more.

Planets
    Earth, 10
    Jupiter, 12
    Mars, 3
    Mercury, 5
    Pluto, 15
    Saturn, 12
        Rings of Saturn, 13
    Uranus, 17
    Venus, 2

Index Punctuation

Index Commas

Use a comma to:

- Separate a word from its page number; do not use *any* punctuation after an index that has subindexes.
- Separate page numbers.

Index Semicolons

Use a semicolon to separate parts of a multiple cross reference. For example:

Automobiles. *See also* **cars; vehicles.**

Index Parentheses

Use parentheses to indicate alternative endings to words. For example:

Europe(an), 129

Index Cross Reference

Use periods to separate an index from a cross-reference. For example:

Automobile. *See also* **cars.**

Index Acronyms

Other items besides words often belong in an index. If your document doesn't have a glossary, it may be helpful to mark in bold font the location of word definitions in the document. If acronyms or abbreviations are used frequently, list them, and indicate a reference to the full phrase. For example:

KB. *See* **kilobytes.**
LAN. *See* **local area network.**

It is also customary to list an index's acronym or abbreviation with the term. For example:

kilobytes (KB)
local area network (LAN)

Index Numbers

If numbers are listed as entries, these should go at the beginning of the index, before entries beginning with the letter "A."

Indexing World Wide Web Conventions

World Wide Web and other web indexes should be alphabetized under their full universal registry locations (URL). For example:

World Wide Web
    web browsers, 243
        ftp://www.culture.canada.org/f12.htm. *See also* Canadian
     Culture.
    http://www.alt.moscow.ru/igor.htm. *See also*
     Russian music.

Index Jargon

If jargon is used in your particular document (and in general, it should be avoided), list the jargon term in the index if you think people would use that term to find something; always remember the perspective of your audience.

In addition, Internet technology has created a vast amount of jargon-like terms that have no easy translation without using additional jargon. In these cases, technology is creating language that walks a fine line between technical terms and jargon, and is creating new words faster than society can absorb them.

For example, a *kill file* is a jargon term for an executable that disables specific chat in an online chat room or newsgroup. But most people understand the term only as *kill file,* and would search for this term in an index.

Determining What to Index

Introduction
This section offers an approach for determining what terms to index. It is intended for those who index using word processing software, and will also be helpful to those who edit indexes generated by online reference systems. At the end of this section is a completely manual approach to indexing with the use of index cards, which is intended for those few writers who have no software indexing features available.

Operate the Index Features
Before writers can index a document, they must first learn how to operate the indexing component of the software. If the software comes with its own indexing styles, use the one that most resembles these conventions or the index standards used in your corporate style guide. Obtain a comfort level with the indexing features (which can often help you to index efficiently), then choose an approach for indexing.

Two Approaches to Indexing
There are two possible approaches for indexing:

1. Create the index simultaneously as the document is written.
2. Create the entire document, edit it, then index it.

Write and Index Simultaneously

In the first approach, writers decide what to index as they write and create the index simultaneously with the document. The advantages to this approach are:

- The index is complete when the document is complete.
- Indexing doesn't seem to be as tedious.

The disadvantages to this approach are:

- The index needs editing after the document is edited.
- Working on the index can interrupt the flow of thought a writer needs to write well.
- If the system is online (or even on some word processing software), changes to the document or its index can create problems with the system pointers used in linking index terms to their references. (Refer to the previous section, *Index Pointers* for more information.)
- There is a tendency for writers to index too many terms in the beginning of the document and not enough terms toward the end of the document.
- Writers may not understand the full impact of the document's content and what should be indexed until it is completed.

Write Index Last

In the second approach, the document is finalized and then indexed. The advantages to this approach are:

- An accurate index can be generated without being re-edited, and system problems with pointers can therefore be avoided.
- Some time has elapsed between writing and indexing, which often offers a more objective perspective on determining what terms to index.
- The finished table of contents can be used as a guide for determining how much weight to give to each topic within the index.

- It is easier to set a limit on the length of the index and be more consistent in indexing when the document is finalized and its length is known.
- Indexing can accommodate a shift in content perspective that may be realized only after the document is complete: this shift in perspective may impact the choice of what terms are indexed.
- Another writer or a subject matter expert could index the document and offer a fresher perspective on what is indexed.

The disadvantages to this approach are:

- Indexing can seem more tedious.
- If there is a tight deadline or a shortage of time, often the index suffers in quality.

Considerations for Deciding What to Index

Indexing is like writing: the more you do it, the better you become at it. If you have never indexed before or have always struggled with it, this section should clarify what is important to index.

Finalize the Table of Contents

After choosing your indexing approach, you need to decide what to index. Most likely you chose the second approach to indexing, that is, finalizing the document before indexing its terms. You should also finalize the table of contents before you index.

A detailed table of contents can guide you in determining:

- What topics should be indexed and cross-referenced.
- How much representation from each chapter should be reflected in the index.
- How long the index should be relative to the length of the document (a good guideline is one page of index for every twenty pages of text).

The table of contents can offer other valuable information as well:

- What are the most valuable sections of the document? What are the least valuable?
- Are the earlier sections of the document introductory or overviews? Is there a large appendix?
- Are there graphs, tables, figures, or pictures that should be noted in the index?

After considering the table of contents, you should have an understanding of what level of attention to give to different sections of the document. If you are using a word processor or an online reference system that offers indexing features, you can now begin to index using those features. First however, consider your audience.

Indexes and Audiences

The single most important thing you can do to produce an excellent index is to keep your audience's perspective in mind. With indexing, it is helpful to understand the different types of readers who will use your index. There are two general categories of readers who use indexes:

- Seekers
- Researchers

Seekers already know that the information they seek is in the document; they use the index to find specific references to that information. Researchers do not know whether the information they want is in the document or not, but they use the index to see if the information is there. They may also use the index to identify specific references to information for a bibliography or other reference list.

Index Considerations

As you index, it would be helpful to review the following considerations on every page of content. If you use these considerations as a check-off

Chapter 6: Indexing

sheet of things to look for and to index *on every page* of text you index, you will end up with an excellent index. Does the page offer:

- A rule, principle, or fact?
- A new term or definition?
- An important term?
- An important idea or concept?
- A conclusion to an idea?
- A solution to a problem?
- An important financial figure?
- A historical fact?
- A sequence of events?
- A summary of events?
- A list?
- A procedure?
- A process?
- An important name, date, or place?
- An important graph, diagram, or illustration?
- A list of advantages or disadvantages to an idea?

Indexing Time

It is obvious that a poor index can be created quickly and a good index will take more time. There are other factors that influence the amount of time indexing takes:

- Writers who are very experienced in indexing will produce indexes faster than inexperienced writers.
- The complexity of the software's indexing features will impact the time it takes to index.
- The system response time may influence the time it takes to index.
- Knowledge of the content influences how long it takes to decide what to index.
- Familiarity with the document's topics and locations can influence how long it takes to decide what to index, and how long it takes to create cross references.

Audience Index Considerations

As you index, remember who your target audience is. What things might they look for? What terms might they use?

One inherent problem with using indexes is that people tend to use different words to describe the same things. For example, one person will look in the index for the word, "automobile," another will search for "car," and another for "vehicle." There are only two things writers can do to alleviate this problem:

- Know your audience and try to guess which terms they would use.
- Write consistently and use the same terms to mean the same things through the document.

Manual Indexing

How to Index Manually

Use the same indexing procedures and considerations previously mentioned to determine what to index. As you go through the document write each index entry, its page number, and each word's associated phrase on *one* index card. Keep one word on one card. Continue this process throughout the document. When you're finished, alphabetize the index cards, and type the index using the conventions and punctuation previously mentioned in this chapter.

Chapter 6: Indexing

Chapter 7:

Paper Delivery

Introduction

This chapter is intended for anyone who will be delivering documentation on paper and who needs information on:

- Determining what kind of information is best delivered on paper.
- Choosing fonts, formats, and graphics.
- The use and purpose of side labels.

The Impact of Paper

Paper has had a remarkable history. The Chinese first began using it in about 2300 BC, and for over two millennia it held the most important information in the world. Information for legal contracts, peace treaties, literature, love letters, and of course paper money (although some paper money is really made of linen), all depended upon paper and the ability to discern what information should be upon it. For most of the twentieth century the world seemed to be ruled by information that was displayed on paper. Most governments and businesses of the world could not have evolved into the entities we know without the use of paper.

End of the Paper Trail

With the advent of personal computers, people began to envision how personal computers might be used to eliminate or reduce the use of paper. The invention of hand-held computers and electronic books may one day indeed eliminate the need for most of the world's paper. One day these devices will be as common as pencils and will hold all of the information a person or a business needs. Until that day however, paper will remain an option for displaying information. How good of an option it is depends largely upon the presentation of the message and how the message will be used.

Advantages of Paper

Paper has distinct advantages: is it light, portable, mutable, recyclable, inexpensive, and comes in many different forms (no pun intended). Paper presentation has a number of advantages that make it very useful and user friendly:

- Paper has a physical presence; the amount of information can be estimated by glancing at the size of the paper.
- Paper does not need electricity to operate.
- Paper is inexpensive.
- Paper requires little maintenance if the information doesn't change.
- Small documents in small number are easily portable.
- Information on paper can be very confidential; the creator of the information can often be assured that only one copy of the information exists, and its creation may not leave a trace.
- The information on paper will not be erased by unexpected electrical charges.
- Readers are familiar with the chronological flow of information that most paper presents and can easily determine if they are in the beginning, middle, or end of the flow of information.

Disadvantages of Paper

With very large documents the many disadvantages of paper manuals become obvious:

- The information on paper cannot be easily changed, and if it is changed it must be reprinted and redistributed.

Chapter 7: Paper Delivery

- Changes made on paper can be difficult to track, and there is a good chance that not all readers access the same version of information.
- Paper cannot be distributed to large numbers of readers easily and quickly.
- Large amounts of paper are heavy, difficult to store and keeping track of different versions can be cumbersome.
- Paper presentation and format cannot be customized for each reader.
- Large manuals can intimidate some readers.
- With paper manuals there are only two ways to search for information, either by using the table of contents or the index.
- Pages can be damaged or missing, and they harbor allergens, mold, paper mites, viruses, and bacteria.

Considerations for Paper Delivery

Whether or not to deliver information on paper depends largely upon the audience and their access to electronic delivery. In considering whether or not to deliver information on paper, consider the following questions:

- Will the information change frequently after it is on paper?
- Can the paper be easily distributed?
- Are there many copies of the paper to be stored?
- Is the information well organized and comparatively short?
- Does the information need to be portable? (Will it be read at home, or discussed around a conference table?)
- Does the information need to be used when there is no electrical power? (For example, a hurricane guide should reside on paper.)

Paper Format

The word *format* refers to the layout of the page and the *design relationship* between the text, graphics, and white space. Format contains all the design elements that make the page look a certain way, and includes:

- Text size, positioning and font (letter style).
- The ratio of white space (or empty space) between the text, graphics, and the margins.
- Size of indentions and margins.
- Arrangement of text and graphics.
- Inclusion of headings, page numbers, disclaimers, and so forth.

Font Size

The choice of *font* (letter style) can influence how readable your paper is and how well it is received. Units called *points* measure fonts. A point is 1/12 of an inch or approximately 1/6 of a centimeter. The smaller the point size the smaller the font size. Remember that the world's population is aging, and due to the natural aging process of the eye, older people have a harder time reading smaller fonts. For paper presentation try not to use a font smaller than 12 points.

Font

The shape of the font also influences readability. Notice the differences in the two paragraphs below:

Serif Fonts

Serif fonts, as shown in this paragraph and throughout most of this book, have serifs. *Serif* means, "stroke," and is derived from the Latin word *scribere,* which means to write. Serifs have little strokes on the ends of the letters, which help to separate each character and to guide the eye from one letter to the next; this makes words appear more as one unit and makes them easier to read.

Sans Serif Fonts

Sans serif fonts, as shown in this paragraph, do not have the strokes to guide the eye. *Sans* is French for *without.* Studies have shown that serif fonts, such as the one used in this book, are easier to read and are preferred by readers. Readers are also more accustomed to seeing serif fonts in books and newspapers, and therefore may also be in the habit of reading serif fonts more quickly.

Choose serif fonts if your information will only be presented on paper; if your information will be presented on both paper and online, then other factors (such as screen resolution) for online presentation should be considered; in those cases, san serif fonts should be considered. Refer to the next chapter, *Online Delivery*, for more information.

White Space

Introduction

White space is the empty space devoid of text or graphics on a page. White space has several functions including:

- Helping to organize the text and graphics on a page.
- Guiding the eye to subtitles, paragraphs and other content changes.
- Helping pertinent information to be more visible. For example, important items in a bulleted list are more visible if there is more space around them than around the rest of the text.

Differences in white space usage are obvious to all of us. For example, inexpensive paperbacks or catalogs have every page crammed with small text and use very little white space. There is little white space in order to sell as much text on as little paper as possible.

In contrast, blueprints, engineering schematics, or process flows use a lot of white space to illustrate relationships and concepts, and to emphasize the information on the page. The more white space there is, the more attention is given to the information on the page. For example, when a bulleted list is on a page, the white space around it draws attention to the bulleted items, and they are more likely to be read. When a lot of white space is used, it is because the value of information is more important that the cost of the paper or printing. Imagine the difficulty in reading a building's blueprint if the information was as cramped as a mail-order catalog page.

Chapter 7: Paper Delivery

White space can even be used to encourage readers to consider only one thought at time. For example, many of you are familiar with inspirational books, gift books, and humor books: many of these books have only a couple of lines written in very large font per page. This type of presentation encourages a reader to consider one thought at a time, as no other thoughts are competing for the reader's attention. This type of presentation also requires more page turning and would not allow for a quick, comfortable, reading pace.

White Space Guidelines

Most technical writers allow a ratio of two-thirds text to one-third white space. This means that approximately one third of the page is blank, and this space is used to direct readers to titles and ideas. Understand that this white space is evenly distributed upon the page. This ratio achieves a balance between readability, white space, and economy of paper. Generally speaking, the more white space you have, the more readable your document will be, and often more attention will be given to its contents.

"One-third white space" may sound like too much: it isn't. Use wide margins, which have the added benefit of allowing a page to be read on a screen without having to scroll from side to side. To achieve approximately one-third white space on a regular size page or screen, use:

- Several paragraphs
- Several subheadings or side labels
- Single or double spaced blank lines between each paragraph and each heading
- Bulleted lists
- Twelve point font
- One and one half to two inch margins (six to eight centimeter margins)

The goal should be to design pages so readers can simply glance at a page and immediately know what information it contains and where upon the page specific information is located. Much of this goal is met by the generous use of bulleted lists and side labels.

Side Labels

Introduction
Side labels are also called *topic labels* or sometimes *paragraph labels*, although writers do not necessarily label each paragraph. Side labels guide readers to specific information quickly and help them to read the page faster.

Side labels were first used in religious texts during the Middle Ages and fell out of fashion for several centuries. They started to be used again within online reference systems in the early 1980s. They are very useful for both online and paper presentation.

Side Labels Help Readers
Side labels are indispensable for showing readers different topics very quickly. The choice of wording in the side labels can summarize the idea in the paragraph, or even help to sell or advertise an idea or concept that it labels. With each topic or paragraph labeled, it is much easier to scan documents quickly for specific information. The purpose of side labels is to:

- Tell readers what the paragraph or section is about.
- Serve as subtitles and to guide the reader to the main idea of the paragraph or topic.
- Aid readers in finding information quickly.

The use of side labels is highly recommended and they have become part of the corporate style guide in many companies. Look at the following example of a page with side labels:

Example of Side Labels

Understand Business Drivers	Before a reuse strategy is developed for any company, it is important to analyze and understand the business drivers. The most common business driver for reuse programs is time-to-market. It is important to determine what changes in application development capabilities are desired as a result of reuse.
Choose a Business Model	Once it is understood what the specific goal of reuse will be, it is helpful to have a business model that describes the ways in which IT resources and business areas interact to do business. This model might include: roles, responsibilities, facilities, products and processes.
Common Reuse Elements	Regardless of what reuse methodologies are used, or the scope of the project, there are common elements necessary for a sustainable reuse program. They are: • Inventory • Catalog • Reuse administrator and facilitator • AD methodology • Design standards and principles • Measurement • Quality assurance • Performance incentives

Side Labels Help Writers

Side labels serve as an organizing tool to check the flow of the content. Writers can edit or change the side labels as they add detail to the content. If a side label is too long or wordy, often the idea being expressed by the paragraph is too long, and may indicate to the writer that the content should be in two or more paragraphs. There are three disadvantages of side labels:

1. Writers must take time to create the columns needed for the side labels and align the side labels appropriately. This effort, however, greatly enhances readability and is generally worth it.

2. Commercial printing of documents with side labels (such as lengthy pamphlets and books) is considerably more expensive than documents without side labels. This is the primary reason *this* book does not use side labels, and why you rarely see them in books. (In fact, the small topic headings you see in this book were originally side labels.) This book's publisher, WhiteFeatherPress.com, Inc., estimated that the use of side labels, and the additional white space they require, would have increased the number of pages by about 30%. This increase would have added about 40% to the cost of printing and to the cost of the book. Buyers like you would have paid that much more for the same content you purchased in this format.

3. Side labels are extremely difficult and time consuming to typeset, as each side label may need to placed individually. This would also add to the cost of producing a book.

Many different word processing software packages allow a writer to generate a table of contents from the side labels themselves. This allows for a very detailed and useful table of contents. To do this, it is important to set the style of the side labels to something the software can recognize in generating the table of contents, such as a subheading. Consult the online help within your software to determine how to do this.

Hints for Using Side Labels

When using word processing software, side labels are easier to create and manage if you create either two columns or a two-column table, depending upon how your software handles columns. Use a narrower left column for the side labels, and a wider right column for the body of the text. This allows the text to be written without using the **TAB** key and allows the text to be aligned automatically.

Paper and Online Delivery

Many times in companies a document will be read online in a word processor, an e-mail, or an online reference system; it may also be printed for reference in a meeting, or to read at home. In these instances it is important to consider both paper and online design in order to make your document easy to read, regardless of how it is accessed. Refer to the next chapter for information on designing documents for online presentation.

Chapter 8:
Online Delivery

Introduction

This chapter is intended for anyone who will be delivering information with an online reference system and needs information on:

- Different types of online reference systems and their uses
- Online cue cards
- Online procedures and processes

Benefits of Online Delivery

Online delivery has many advantages over paper, and with electronic books, online delivery may become the dominant medium for delivering information in this century. Common benefits of online delivery include the following:

- Online information can be updated easily and very quickly.
- Online information can be distributed to a large audience almost instantaneously.
- There are no paper manuals to print, store, transport, or distribute; therefore, online information has consistently been shown to be very cost effective.
- Users do not have to insert pages, replace portions of manuals, or throw away old manuals; therefore, they access consistent information.
- Information access is faster than searching through a paper manual, thereby saving considerable money in a production environment.

Chapter 8: Online Delivery

Online Delivery Impacts Information

By the late 1980s the PC and its technological advances had revolutionized the IT industry. There was a paradigm shift from "data processing" (DP) to "information technology" (IT), which reflected the conceptual shift from data as a separate entity to data as meaningful business information. Data became a corporate asset, and many companies had the foresight to implement online reference systems, which allowed them to utilize corporate information to gain a competitive advantage.

As PCs invaded the common business office and the line between the technical professional and the business professional began to dissolve, online reference systems became the foundation for sharing information quickly. This enhanced customer service and harnessed the knowledge within the company. The newer online reference systems have created distinct categories that are discussed in the following section.

Different Types of Online Systems

Introduction

It is helpful to have an understanding of the various types of online reference before deciding if online presentation is suitable for your documentation and the needs of your readers. There are several major types of online reference this chapter addresses:

- General online reference
- Cue cards
- Help systems

Online Reference Definition

Online reference refers to electronically-displayed documentation systems that house text for reference purposes. Usually these systems contain very large volumes of manuals that are frequently updated and accessed by many users. These systems can be mainframe-based, client-server-based, or Web-based. Online reference systems were the first documentation systems to be developed, so all other online systems are a subset of this major category.

Chapter 8: Online Delivery

Types of Online Reference Information

An online reference system should be considered for the following types of information:

- Large amounts of textual information, such as user manuals, reference guides, policy manuals, procedure manuals
- Documentation that changes often
- Sets of interrelated documentation
- Documentation for the handicapped or visually impaired
- Engineering or scientific specifications
- Diagrams or process flows
- Standards manuals, long procedures
- Directories or catalogs

Cue Cards

Cue cards are a specialized type of online reference, and may be a separate system or part of a larger online reference system. Cue cards give very short references or instructions on very specific information; usually each cue card contains a short paragraph of information and may contain graphics or pictures. Cue cards are used to give just-in-time instruction to readers. These systems can also be mainframe-based, client-server-based, or Web-based, and are usually accessed by key-word searches. Cue card systems are used in many industries primarily as "job aids" for highly specific tasks. Cue cards can offer:

- Customer service representatives specific answers to questions telephone customers may ask.
- Engineers answers to specific design questions.
- Shoppers information on what products are in a mall when they key in questions from a mall kiosk.
- Computer programmers help on a development question.
- Computer operators instructions on how to perform an Initial Program Load.
- Factory workers specific instructions on how to handle different phases of a production process.

Chapter 8: Online Delivery

Types of Cue Card Information

A cue card system should be considered for the following types of information:

- Procedures for completing specific tasks
- How-to information for helping customers
- How-to information for processing a component of work
- Short answers to specific questions, such as word definitions

Help Systems/Help Authoring Tools

As the name implies, help systems or help authoring software assist readers using a software application or tool. These applications may be written for programming tools, online reference tools, computer-aided drafting tools, business process tools, document management tools, Web-based tools, or virtually any software application there is.

Users access the help system by pressing a function key or clicking on a help icon. The user can then search a help index or type in key words to find the information they need. Usually help systems are written independently of the associated software application and are then imported into the application. The application should already have space allocated and a file format that is compatible with the help system. Occasionally the help-authoring tool is built into the application, and writers merely have to write the help text within the application itself.

Types of Help Systems Information

The type of information usually accessed from a help system is very short procedures for using a software application. This information is very specific to tasks the user is trying to complete and may contain a short, one-to-five-task procedure. However, in the late 1990s, help systems became very sophisticated and the type of information they offer expanded; they are now used for virtually every kind of help any user on any system with any computer may need. Help systems and cue card

Chapter 8: Online Delivery

systems are beginning to overlap the types of information they contain, and the line between them is becoming increasingly blurred. What remains important is the quality of information offered, and not the name or type of the system in which the information resides.

Computer-Based Training

Computer-based training (CBT) is another major category of online reference, though it is not discussed in detail in this book. A basic CBT system consists of online reference software that contains a course. Users enter answers to course questions (often in a multiple-choice format), the system tracks the user's responses, and scores the answers.

More sophisticated CBT systems are highly intuitive, meaning that they adjust the level of difficulty in the questions according to the success rate of previous answers the user has given. They may also allow users to query the system or write essay answers to questions. The special requirements of CBT (an understanding of adult learning principles, curriculum design, and interactive reference software) are beyond the scope of this book.

All of these online reference systems can serve different purposes, and they can work together in the same environment to give users just-in-time information they need to perform their jobs. However, the best systems in the world won't help users perform better on their jobs if the information is organized poorly or if it is difficult to access. Read the next section to understand the basics of organizing information in online reference systems.

Chapter 8: Online Delivery

Organizing Online Information

Introduction

Refer to this section for information on how to determine a system structure for storing, maintaining, and accessing information in an online reference system.

System Structure Definition

A system structure is the high-level design of how online or Web topics will be stored, maintained, and accessed from an online system. A system structure consists of a navigation map that identifies system links and the paths of their interconnections. Choosing the right system structure will assist in maintenance and enable users to find information more quickly.

System Structure Purpose

The purpose of a system structure is to organize the links between topics; this is not the same as organizing the content within topics, which is discussed later. Users may or may not have access to the map itself as a navigational tool, but nonetheless it serves as a navigational map for the writers, designers, and most importantly, the people who maintain it.

People who maintain the system will depend upon the structure as:

- A map of the links
- A way to track how the links connect
- A way to understand the sequence of topics
- A way to understand how users will access categories of information

Types of System Structures

There are five common types of system structures:

- Sequential Structure
- Hierarchy Structure

Chapter 8: Online Delivery

- Fishbone Structure
- Grid Structure
- Web Structure

Sequential Structure

A sequential structure arranges information in a logical sequence and is the simplest structure possible. In this structure, topics follow one another in an ordered fashion and the user has the limited options of going forward, backward, or skipping around to unrelated topics. The sequential structure is illustrated below and is often used for:

- Catalogs
- Reference code books
- Help topics arranged in alphabetical order
- Other listings arranged alphabetically or numerically
- Computer-based training courses where each topic builds the foundation of knowledge required for subsequent topics

Sequential Structure
(Figure 1)

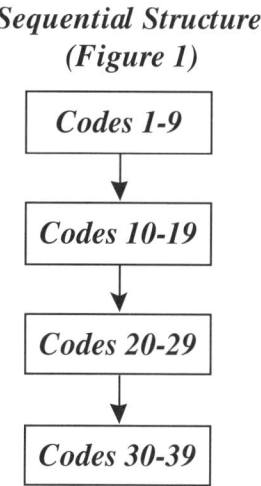

Hierarchy Structure

A hierarchy structure is comprised of various levels of information that begins with a high-level general topic and breaks down into lower-level specific topics. Each level of topic offers more specific and detailed information; in online jargon, this is called *drilling down* to more *granular* information.

The following example illustrates how a hierarchy structure drills down to a more granular layer of information. The top line of the hierarchy is the highest-level topic, and each subsequent topic drills down the subject into more granularity.

Hierarchical Structure
(Figure 2)

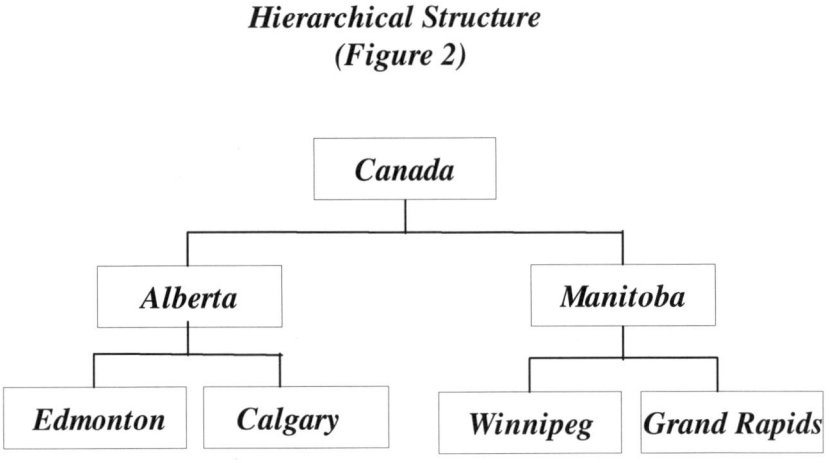

This structure is easy for most users to navigate, but can be difficult to update without impacting or changing the levels of hierarchy. This hierarchy, or fishbone structure, is useful for information that has many categories with sub-topics. When the hierarchy must be modified or updated, try adding a new topic (if it makes sense), rather than rearranging information in the old hierarchies.

Fishbone Structure

A horizontal hierarchy structure is called a fishbone structure, as it resembles a fish's skeleton. It is often used for Web site maps. Some people think that users prefer to navigate by scrolling to the right, as opposed to scrolling downward with the hierarchy. This preference might have originated because most web development software uses right-hand scrolling to create navigational maps used by developers and those who maintain the site. The following graphic illustrates the fishbone structure.

Fishbone Structure
(Figure 3)

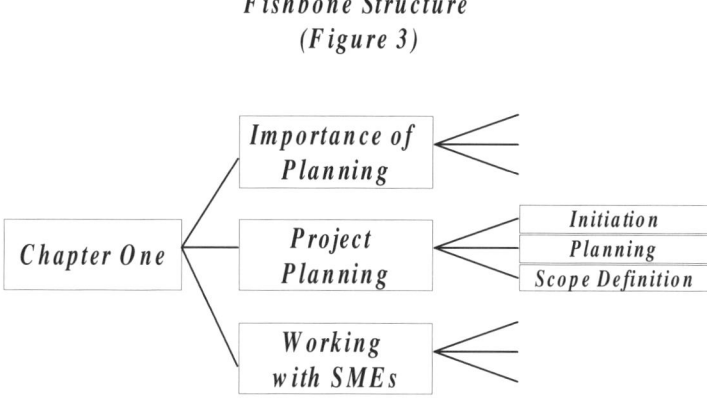

Grid Structure

The grid structure has a long history that begins ages before computers were invented. Grids that display information have been used to display transportation schedules, mathematical charts (such as interest rate charts), formulas, and conversion charts. Grids can display any other type of information that contains two interrelated elements that can be displayed vertically and horizontally. Because of their familiarity, grids are very easy to navigate, which is their main benefit. However, depending upon their initial structure, they may not be easy to update if the type of information to be displayed changes drastically.

In addition, grid structures become difficult to read and to navigate when they are too large. How big is too big? When displaying the navigational map, users should be able to see the entire grid on one screen. Having to scroll past the screen edges to see different portions of the grid causes users to lose their frame of reference and to become confused. For these reasons, grids are being used less and less for online or Web presentation. Grids are more useful as a table of contents for similar categories, as in the following example.

Grid Structure
Figure 4

Mainframe	*UNIX*	*Server*	*Web*
Development Tools	*Development Tools*	*Development Tools*	*Development Tools*
Developer Standards	*Developer Standards*	*Developer Standards*	*Developer Standards*
Development Environment	*Development Environment*	*Development Environment*	*Development Environment*
Database Tools	*Database Tools*	*Database Tools*	*Database Tools*

Chapter 8: Online Delivery

Web Structure

The web structure is becoming more familiar as corporate web sites and World Wide Web sites proliferate, and for good reason. With a web structure, every item can connect to every other item and users have a variety of directions in which to go. Web structures offer many creative options, are easy to update, and can be expanded indefinitely. A disadvantage of web structures is that they must be carefully maintained and each link must be tracked. Any changes in information can destroy links. The following graphic illustrates the web structure:

Web Structure
Figure 5

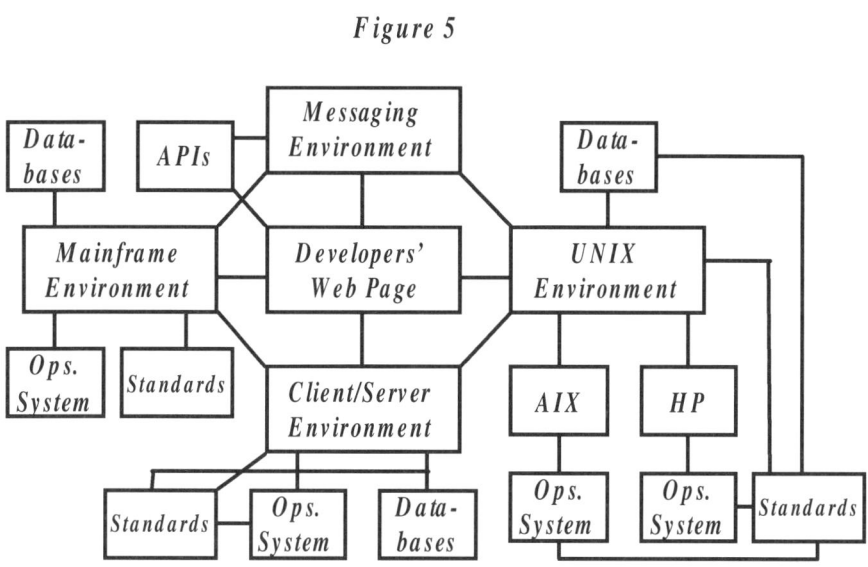

Maintain Links

Maintaining all the links in any structure, however, can be challenging, and the number of links generated by a web structure can be surprisingly large because many of the topics are linked to many other topics, and they may not follow a pattern. Links in web systems are easy to misplace or break. One disadvantage of having many links is that they can compromise system performance.

Chapter 8: Online Delivery

If every topic is linked to every other topic, the number of links rise *exponentially* as more topics are added. For example:

- With 2 topics there is 1 link
- With 3 topics there are 3 links
- With 5 topics there are 10 links
- With 10 topics there are 45 links
- With 20 topics there are 190 links
- With 100 topics there are 4,950 links

Formula for Determining Number of Links

The following is an easy algebraic formula to determine the number of links for a given number of topics. This formula is used in many industries to determine the number of links between multiple points. In this formula *T* equals the number of topics, and *L* equals the number links. In the following example, twenty-five topics are all linked to each other, yielding three hundred links to be maintained:

$$\frac{(T-1)(T)}{2} = L$$

Example: $\dfrac{(25-1)(25)}{2} = \dfrac{(24)(25)}{2} = \dfrac{600}{2} = 300$

Multiple Structures

It is important to remember that although you are encouraged to organize all of your information into one structure, you may use *two* structures for the same online reference volume or Web site, especially if you have two distinct sets or types of information.

For example, you may use two fishbone structures for two different volumes of information in an online reference system. You could display both fishbone structures in the users' navigational map, which would tell users in which volume a specific topic was located.

If you have two distinct types of information, do not hesitate to use two different structures if those structures would enhance navigation for users and be easier to maintain. However, try to avoid using more than two types of structures for any one online reference volume; this can be confusing to maintain and difficult to navigate. If you are tempted to use more than two structures, it is possible that the chosen structures are not appropriate for your types of information, and that a more versatile structure (such as the web structure) would be a better choice.

Tips for Choosing an Online Structure

An online reference system is useless if it is too difficult to navigate or too confusing, or if the search method is slow or difficult to use. If users get lost for any of these reasons, they will probably not return to the online reference system. Therefore, the system must be easy to navigate and return the correct search results quickly.

Consider the information from the viewpoint of your audience. The following are some tips for a choosing a system structure:

- How will users access and use the system?
- What structure will be logical for users?
- Will users be reading long passages of information or short ones?
- Will users be browsing the system and needing cross references? (If so, then consider web or grid structures.)
- Will users need to find facts quickly, perhaps in a cue card system? (If so, any structure would work that can be quickly accessed and updated.)
- How will the users navigate?
- What are the users' past experiences with online or Web reference?
- Will users receive training on the system?

Chapter 8: Online Delivery

- Will users access the system in just a portion (or window) of their screen while viewing other information in another portion of the screen? Should your online presentation be shaped to accommodate the size of the screen to minimize scrolling?
- Does your system allow users to submit feedback about their experiences using the online system? Will someone in your organization be responsible for responding to inquiries and concerns?
- Should users have the ability to view and navigate using the map structure? Would this benefit or confuse them?

A Note on Audience Considerations

It is highly recommended that you review the section *Gathering Audience Information* in *Chapter 1: Planning*. The best writing in the world created on the best software will be wasted if it is not written with the audience's needs in mind.

Consider System Structure Maintenance

There are many considerations for choosing a system structure because of how they organize links. When choosing a system structure, consider how the system will be maintained:

- Are there dedicated, experienced staff members to maintain the documentation and test the system links? Or will professionals inexperienced with online systems handle these duties in addition to their regular duties?
- How many links will be too many to maintain for the number of people maintaining them?
- What is the system's link limit?
- What process will be used to maintain and track the links?
- What system structures already exist in your organization? How easy are they to navigate and to maintain? What is the organization's experience with specific structures?

- Who will have the system rights to change links? Who should coordinate link changes between organizations?
- How will a link problem be made known? Are users expected to report broken links? Is there an easy process for this?
- Is the system able to overlay links automatically? How will this be assured or tested?

Beware of Vendor Hype

Often the system vendors will tell organizations that there are no limits to the number of links, and in theory this may be so. However, there are always physical limitations on the network system as to the number of links it can handle. Too many links for a system will cause poor response time or more serious system problems.

Beware of vendor hype. Do not be afraid to ask the vendor specific questions and demand specific answers. Be certain that you understand all the possible conditions of when and how the software executes upon a system. Be certain that the technical resources in your organization also understand these factors and are comfortable with the software and its performance load upon the system.

If you are considering purchasing an online reference system, install and test the system in your own environment. Be aware that vendors develop and test their products in environments that are optimal for their products. How the product acts in *your* environment may be quite different.

Chapter 8: Online Delivery

Types of Online Information

Cue Cards Definition

Cue cards are short passages written to give just-in-time information to perform a task. Cue cards are always short (100 words or less) and give users short answers to questions they may have; in this way cue cards act like a help system, as users can access them by entering key words. Cue cards are also used to break down *procedural* information into small steps that can be accessed quickly.

Cue Card Example 1

Abend Code SOC 7:

The final two slashes and three asterisks (//***) were missing from the last DD statement of the JCL stream.

Correction:
    1. Access the application's JCL.
    2. Insert //*** in the last DD statement.
    3. Resubmit the job.

The topic of the cue card is very specific; it does not offer definitions for a range of abnormal ending (abend) codes, but only a specific abend code. The cue card is short (41 words), and it is worded succinctly. There are only three steps to correct the problem.

If the steps were longer, the other steps could appear on a second card. Try to keep cue card procedures shorter than seven steps; more steps may indicate that you have really have two or more procedures that should be accessed on their own specific cards.

The organization of the cue card is visually well designed. The answer to the user's query is provided on top, and the steps to correct the problem are located in the center of the card and set off from the rest of the text. In this way, hurried users have the option of correcting the problem without reading what the code means.

Cue Card Example 2

008 Server Access Denied:

We are sorry you are not authorized to access this server, which contains proprietary corporate information.

Contact the Corporate Legal LAN administrator for permission to access this server.

The previous cue card has 29 words, and explains the meaning of a particular error code and how the user can correct it with a one-step procedure. Notice the simplicity of this card: it contains three sections that are visually set apart. The first section shows the error code and might be considered the title of the card. The second section shows what the code means. The third section tells the user what to do. No extra information is given.

Cue Card System Structures

Cue cards work well with the following system structures:

- Sequential (is excellent for a range of cards that users may want to access in order)
- Hierarchy (works well for cue cards that have different levels of information)
- Fishbone (works well for cue cards that have different levels of information)
- Web (works well for cue cards that are not related by order or level of information)

Chapter 8: Online Delivery

Online Procedures

Procedures are specific action steps to perform a task that may be delivered in cue cards, or a long procedure: they tell *how* to do something. Procedures have numbered steps to be performed in an order, and offer a very detailed explanation of how to perform a specific task. They are usually titled with the *ing* form of the verb. For example, *Researching Error Codes*. They can also be titled with *How to* . . . .

Short procedures work well as one or more cue cards. Longer procedures or complex procedures that need explanations of the steps should *not* be delivered as cue cards. Following are two examples that illustrate these differences; the first example is a cue card and the second example is not a cue card.

Procedure Example 1

Resetting User Passwords:
1. **Ask the user for the RACF ID.**
2. **Press F4 to access the User Profile Listing.**
3. **Find the RACF ID.**
4. **Find the associated level of authorization in the third column of the User Profile Listing. (If the user is not authorized, ENTER keywords "user not authorized" to access that procedure.)**
5. **Press F2 to access the Change Password Screen.**
6. **Change the password to a temporary password.**
7. **Tell the password to the user and indicate that it must be changed within three minutes.**

This cue card has seven steps (which is the longest a procedure on one cue card should be) and has 101 words (which is the upper limit for one cue card). Note the diversion in step four: the author may have been tempted to have the "user not authorized" procedure embedded in this procedure, but wisely chose not to do that. Embedding another procedure

Chapter 8: Online Delivery

would have made the cue card too long, and would have broadened the scope of the cue card to something that was less specific.

If your procedure is more complicated or requires more detail, a cue card format may not be the best. The following example uses the same procedure, but offers much more detail. It is not delivered as a cue card, but is part of several screens of text.

Procedure Example 2

Resetting User Passwords:

Use this procedure to change a user's mainframe password.

1. Ask the user for the RACF ID.
2. Press F4 to access the User Profile Listing.
3. Find the RACF ID in the first column of the User Profile Listing; the IDs are listed in alphabetical order.
4. Find the associated level of authorization in the third column of the User Profile Listing.

If the user is authorized:

1. Press F2 to access the Change Password Screen.
2. Change the password to a temporary password. Press ENTER.
3. Tell the password to the user and indicate that it must be changed within three minutes.

If the user is not authorized:

1. Ask:
- Why access is needed.
- Who is the user's LAN administrator.
2. Obtain the user's phone number, explain that you will arrange access and will call back when access is granted.

Chapter 8: Online Delivery

3. Call the LAN administrator and explain the problem.
4. Wait for the LAN administrator's verification before granting access.
5. When access is granted, press F2 to access Change Password Screen.
6. Change the password.
7. Call the user with the new password.

This procedure is called an *if/then* procedure, meaning that different series of steps are performed for different situations. If the procedure is an if/then procedure, and if cue cards are used, the different procedures would be located on different cards and linked together for easy cross reference.

Procedure System Structures

Procedures written in cue card format work well with the following system structures:

- Sequential (is excellent for a range of cards that users may want to access in order)
- Hierarchy (works well for cue cards that have different levels of information)
- Fishbone (works well for cue cards that have different levels of information)
- Web (works well for cue cards that are not related by order or level of information)

Procedures that are too long or complicated for cue cards work well with the following system structures:

- Sequential (is a natural structure for complicated procedures, and probably the best choice)
- Hierarchy (works well for some procedures that have many subprocedures)

Chapter 8: Online Delivery

- Fishbone (works well for some procedures that have many subprocedures)
- Grid (works well for several procedures that are closely linked with common characteristics: for example, if there were three procedures for "accessing systems," the common characteristic linking the procedures might be three different systems)

Procedures Versus Processes

Procedures and processes are very often confused even though they are two distinct types of information. Procedures tell the "how-to" of a situation and have action to be performed within a relatively short period of time. Procedures are usually written in second person, or the *you* form of speaking.

Processes summarize how to resolve a situation and give a high-level understanding of how a series of different procedures are linked together to achieve a certain result. Usually processes are written in third person, (the *he, she,* and *it)* form of speaking.

Online Processes

A process is series of events that achieve a result. Processes summarize the different procedures existing within the process, but the details and steps of the procedures are not important for the type of information the process conveys. A process is the *big picture* and often involves different people with different responsibilities.

The following example is the *process* for the examples of procedures previously presented; notice how the process differs from the procedures.

175

Chapter 8: Online Delivery

Process Example

Password Change Process:

When users need their passwords reset they call the Help Desk, who can often reset their passwords while users are on the phone. If the Help Desk cannot reset the passwords, the LAN administrator must authorize access for the users before the Help Desk can reset passwords. Depending upon the administrator's availability, this may take a full business day.

<u>Help Desk Procedures</u>
<u>LAN Administrator's Procedures</u>

Note that the process gives the "big picture" of how passwords are changed; it does not say how to reset the password, but explains the different people involved and what their responsibilities are. The time span of the process is relatively long. Processes are an excellent way to explain how events and organizations are linked to achieve a certain result, whereas procedures explain how to accomplish a specific task.

Processes and Readers

Processes are usually too long and complex to work well in a cue card format. However, if a short process must be presented in a cue card format, be sure to tell readers where they can find the procedures. In the process example above, the cue card has underlined hyperlinks to two sets of procedures. If a process is on a cue card, be sure to link the card to procedure cards too. Usually processes are presented in a separate section for the online manual or corporate web site. Processes often work well when presented *before* procedures to explain to readers:

- What their role is in performing a procedure
- Who else is involved in achieving the desired outcome
- Where the work originates and where it ends

Process System Structures

When processes accompany procedures, they can use the same system structure as the procedure. When processes are presented separately from procedures, they work well with any of the following system structures:

- Sequential (is excellent for multiple processes that may be sequential or linked in a certain order)
- Hierarchical (works well for multiple processes that may be grouped together and that have sub-processes or sub-procedures)
- Fishbone (works well for multiple processes that may be grouped together and that have sub-processes or sub-procedures)
- Web (works well for a different array of processes that have no common characteristics)

General Reference

General reference can contain all types of information except cue cards, help systems, and catalogs; it is the type of information most people think of when they think of "online reference" or "Web reference." General reference often contains information that has multiple purposes and that may be accessed by different types of users seeking different information. If your documentation has multiple purposes, try to choose a system structure that will work well with all types of information.

Chapter 8: Online Delivery

Chapter 9:
Screen Design Considerations

Introduction

This chapter offers considerations for designing online reference and should be read before any documentation is placed into an online system. The following basic guidelines for designing online reference are accepted throughout the technical writing industry.

Screens Are Not Books

Obviously screens are not books, but there are differences you may not consciously realize. Books have a physical presence; users can see a book and instantly judge its size. Users can open a book on any page and have an idea of whether they are near the beginning, middle, or end of the book. In this respect, online reference systems have no physical presence, and the design of the online screen must help users navigate through the system. Therefore, consider how your documentation will be displayed upon the screen before you begin writing it.

A Screen Is Not A Page

A screen is not a page; it can display approximately one-third of the information that can be displayed upon a page. If the screen is filled with text, it can display a maximum of about 300 words.

Only about 200 words or less per screen are recommended in a good design that has the right amount of white space, or blank space.

Chapter 9: Screen Design Considerations

Screen Layout

It highly recommended that a standard screen layout be used. Allocate the same amount of space for the text, white space, and graphics on each screen, and keep it consistent, so the text and graphics do not appear to "jump around" from screen to screen. In assigning screen space, you may want to use boundaries comprised of blank space, boxes, or borders to set off text and graphics. Use generous boundaries because these can "shrink" depending upon a user's display settings and screen size. Small boundaries can cut off information on some screens.

Screen Space

Screen space can be divided into areas for specific uses. Often screens are vertically divided into columns to indicate subtitles or side labels, and to help users visually group associated text.

Screen Navigation

Use labeled arrows or buttons to indicate "next" screen and "previous" screen to help users navigate between screens. If users enter the middle of a topic in an online reference system, they may not be aware that they entered in the middle. Therefore, use the word "continued" at the top left of the screen to indicate that the topic began on a previous screen. You may even want to indicate manual title and chapter on each screen, especially if the manual is one of a set of interrelated volumes.

Screen Shapes

The screen is shaped differently from a page; it is wider and shorter, so information laid out for a page may look tiny with huge margins when directly transferred to a screen: or, users may have to scroll from left to right in order to read a complete line. Users' monitors within the same organization can vary in size and their display settings (which determine the size of the fonts and icons, and the clarity of images) on their monitors can be set differently. Therefore, it is important not to crowd margins so that no user is forced to scroll to the edge to see lines of text or to change monitor settings.

Chapter 9: Screen Design Considerations

Multiple Screens

Many times users will access online reference in a window of their screens while accessing other information in other windows of their screens. If your online or Web reference system will be used simultaneously with other systems, the shape of information on your online screen will need to be adjusted accordingly. Think about how your documentation will be used in conjunction with other references. For example, you may decide that a long, narrow shape is best for your documentation so it can be displayed as a vertical window on user screens.

Screen and Paper Page Numbering

Pages cannot be transferred to screens with the same page numbering retained on the screen. This is because a screen can contain only about one-third of the information on a page, or in other words, it takes three screens to display the information on one page. This could result in confusion if page numbers are not removed before the documentation is transferred online. For example, for every two pages viewed, readers will have viewed six screens, but the page number displayed will still be *page 2*. This is another compelling reason to redesign the documentation for online viewing.

Designing for Online and Paper Use

Documentation designed for paper is very difficult to use online, while documentation designed for online use is still easy to use on paper. If you are unsure whether or not your document will be used online or on paper, design it for online use. Well-designed online documents work well on paper, but well-designed paper documents are often too wide or spaced poorly for online presentation.

Screen Resolution

Screens have much poorer clarity (or resolution) than a page. (Resolution is the space between pixels on a screen and is measured by the number of dots per square inch.) Most monitors have at least 100 dots per square

Chapter 9: Screen Design Considerations

inch, most office print has at least 600 dots per square inch, and books have more than 1200 dots per square inch. This means everything is blurrier on a screen than on a page. Poor screen resolution can tire the eyes quickly.

In addition, often users' screens are in poor light, have glare, or are at a poor angle or distance. Most users read from a distance of about 20 inches, or 40 centimeters. Keep these factors in mind when designing online reference screens, and avoid fonts or images that are too small or fuzzy.

Screen Viewing Areas

Use Consistent Screen Viewing Areas

A screen viewing area is a designated portion of the screen that is consistently used for specific types of information. There are several benefits with using consistent screen viewing areas for an online reference system or Web page. Consistent screen viewing areas save computer memory, as the system can update changes by overwriting only the changed information. Users benefit from consistent screen viewing areas by learning what to expect from each screen; consistency allows users to find information on a screen quickly without having to search each screen. Reading is quicker as users know where to find the information they need.

A good example of this consistency is the practice of listing the contents of a Web site down the left side of the screen; this has become a Web design standard. When users change Web sites they expect to see the contents listed on the left side of their screens, site after site.

Several Types of Screen Viewing Areas

It is common in online reference systems or Web screens to have between two and six screen viewing areas, or frames, per screen. The number of screen viewing areas depends upon the design, the complexity of the information, and amount of information.

Chapter 9: Screen Design Considerations

The following screen viewing areas are listed in order of importance:

- Top margin area
- Primary area
- Bottom margin area
- Content listing area
- Secondary area
- Third area

Top Margin Area

The top margin area or top margin frame is required in any online reference system or Web page and is located across the top of the screen. Its function is to tell users where they are in the system; it serves as the main navigational cue and often displays:

- Document or Web page title
- Subtitles
- Volume number
- Topic titles
- Topic continuation notes (indicating that the topic began on a previous screen)

Primary Area

The primary area or primary frame is also required on any online reference system or Web page; it displays the main information of the screen. The information can be as mundane as "http error: file not found" or as elaborate as an interactive graphic in an online reference system. The primary area is located in the center of the screen.

Bottom Margin Area

The bottom margin area or bottom margin frame is the second navigational area of the screen and is located along the bottom of the screen. Its function is to tell users where they can go next. The bottom margin area is not required, but it is highly recommended, and its usage is an accepted standard in both online reference and Web design. The bottom margin area often contains:

- Links to other locations or e-mail links
- Search boxes (for entering keyword searches)
- Screen numbers
- Copyright information
- "Next" buttons that allow users to continue
- "Back" buttons that allow users to return to the previous screen
- "Continued on next screen" notes for online reference topics that continue on the next screen

Content Listing Area

The content listing area or content frame, is an optional area, but is highly recommended. It gives users additional navigational choices, and it resides in a vertical strip on the left side of the screen. It is also called *navigation area*. It allows users to go to other topics listed in this area. The contents are hyperlinked words, which allow users to access other topics of information.

Secondary Area

A secondary area or secondary frame is an optional area used to display additional information that supports the information in the primary area. It resides in a vertical strip on the right side of the screen. For example, it may contain an illustration or photo of whatever is discussed in the primary area. Examples of items that may be displayed in a secondary area include:

- Help information
- Definitions of terms
- Engineering schematics
- Graphics or photos
- Editorial notes

Chapter 9: Screen Design Considerations

Third Area

A third area or third frame is not used very often, and is typically reserved for graphics, photos, or advertisement. Often this area's designation is small and resides either on the bottom of the secondary area or overlapping the primary and secondary areas.

In spite of the creativity that individual online reference developers and Web developers have, screen designs have defaulted to a few designs that offer users information consistency and predictability. You will find these few designs on about 90% of the world's Web sites. The following graphs illustrate the screen viewing areas previously discussed as they apply to the common screen designs used in online reference and Web development.

Example 1: Common Screen Design
Figure 6

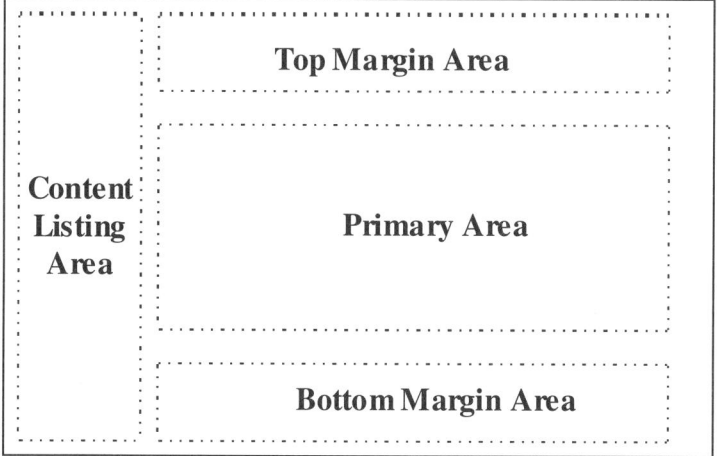

Chapter 9: Screen Design Considerations

Example 2: Common Screen Design
Figure 7

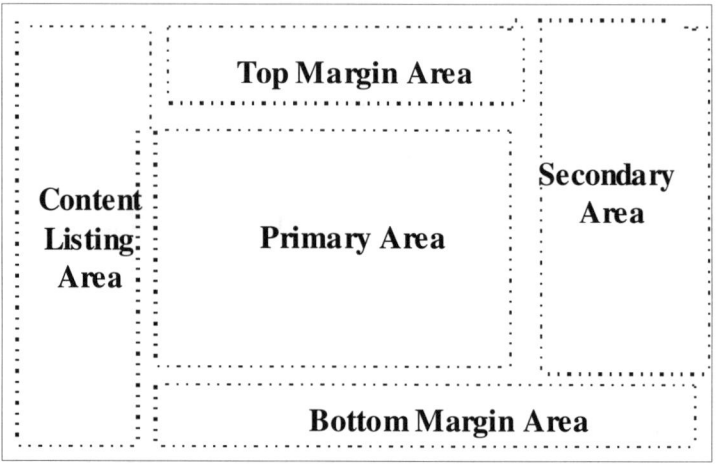

Example 3: Common Screen Design
Figure 8

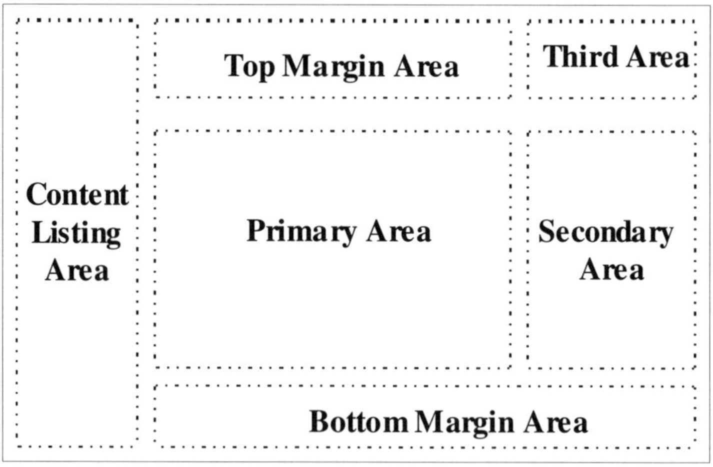

Chapter 9: Screen Design Considerations

Good Screen Design Considerations
After choosing your screen layout, consider these other guidelines when designing your online reference screen:

- Most of the world languages are read from left to right; therefore, place items that users should read first on the top-left side, and items users will read last on the bottom-right side.

- Most users are right-handed and keep the mouse on the right side of their PCs. Put buttons and arrows on the right side of the screen so users have a smaller distance to reach with the mice.

- Use an uncluttered screen background in a light color, such as pale blue, white, light gray, or tan, that allows for contrast between the text and the screen. Bright or dark colors are popular on Web pages, but they tire the eyes if viewed too long.

- Use dark-colored text, such as navy or black, and avoid too much red on the screen.

- Use a very generous amount of white space (or blank space) around the text and graphics.

- If the same topic continues onto more than one screen, write the word "continued" at the top left of the screen so users know they are not at the beginning of the topic.

- Use a consistent screen design for all documentation so users learn where to look for back buttons or navigational clues.

- If you know that all users have very good screen resolution, use serif fonts (fonts with "tails" like this book uses) when possible, as these are easier to read. If you are unsure of how good the users' screen resolution is, use sans serif fonts without "feet" (as in this phrase) to avoid blurry-looking text. Read the following guidelines on fonts for more information on serif and sans serif fonts.

187

Online Font Guidelines

Font Size

For online design, most fonts should never be smaller than 10-14 points. There are some fonts that are still very small for this point size. Try your chosen font at different distances from the monitor; get other people's opinions of what size font is easy for them to read, *then* enlarge it 1-2 points. Reading fonts online larger than 18 points for a long length of time is also tiring on the eyes, so use larger fonts only for titles, subtitles, and other types of headings.

Font Size Guidelines

In choosing font sizes, choose one size for the body of the text, and choose other sizes to contrast with that of the body of the text and to emphasize titles. Use no more than three or four different sizes of font.

A guideline for good contrast is to consider the font for the body of the text as the smallest font, and to choose font sizes that are at a *minimum* one-quarter larger for subtitles and titles. For example, if the body of the text is in 12-point type, make the subtitles at least a 15-point type, if not larger. You may prefer to make the main title font quite large, maybe doubling the size of the body text font.

Font Style

In the early days of online reference, screens had much poorer resolution than they do today, and the use of sans serif fonts was highly encouraged for online use. An example of sans serif font is this sentence; sans serif fonts do not have the "tails" that guide the eye to the next letter. Many people prefer sans serif fonts for online viewing, as they may appear less fuzzy than serif fonts, but sans serif fonts often have thinner lines that do not show well under poor resolution.

The best way to decide which style of font to use is to test them online; have several people view each font from several distances and simply vote on which to use. You may already have corporate standards that dictate which type of font to use under different conditions.

Italicized Fonts

Due to the resolution on even the best monitors, italics are difficult to read online. In addition, they make the lines of each letter thinner and harder to read. Use italics only for emphasis, foreign or legal phrases, and titles. You may want to consider using bold italics, as the lines are thicker and a little easier to read.

Bold Font

Bolding fonts are an excellent way to add emphasis and improve readability on a screen, as bolding increases screen contrast and resolution. Some styles of fonts even look too thin unless they are bold. Although you are encouraged to use bolding, do not use too much bolding; anything that is overdone will lose its emphasis. You may want to consider bolding:

- Titles and subtitles
- Emphasized words
- Screen numbers
- Instructions
- Italicized fonts

Capitalizing Fonts

Avoid using all capital (uppercase) letters for emphasis. When all capitals are used the text is more difficult to read quickly. Users are accustomed to reading a mixture of uppercase and lowercase letters. The varying letter heights with mixed cases give a distinct shape to each word, and these shapes of words are reinforced over a reader's lifetime.

Words in a familiar shape are easier to read more quickly, and reading is almost automatic. If using all capital letters removes this shape, the reader's brain must now read each letter to form the word, and this takes more time.

Chapter 9: Screen Design Considerations

Using Capitals

Therefore, it is recommended that capitals be used under three conditions:

- For occasional use in titles or where emphasis is required; for example, in user WARNINGS.
- To indicate specific keys on a keyboard that users should press, as in "Press ENTER."
- To indicate specific command lines that users must type in, as in "Type in DELETE SYSEDIT, then press ENTER."

Font Underlining

In the days of typewriters, underlining was the only way to emphasize text or to indicate italics. Times change; underlining is no longer used except to indicate a hyperlink, and even then it is being used less and less frequently. Underlining interferes with the readability of whatever is underlined. Underlining isn't even used anymore (on paper or online) to indicate book titles. Underlining crowds the text on the page or screen and contributes to a screen's poor resolution. With PCs there are better ways to emphasize text using different fonts, colors, and styles.

Font Color

Choose font colors that will contrast well with the color of the screen. The best contrast is achieved with a white screen and black font. However, this can be considered boring, so other dark colors such as navy blue can be used. Don't use yellow for fonts, unless it is for a Web page that has a dark background; yellow is difficult to read against a light background. Use bright colors (such as red and orange) sparingly, as they tire the eyes quickly. Reserve bright colors for:

- Shadows on titles
- Warnings
- Graphics

Chapter 9: Screen Design Considerations

Font Blinking

Most people consider blinking font to be very annoying, and it is tiring on the eyes. Avoid blinking font altogether, or reserve it for use in a very special, very dire warning. For example, the word "WARNING!" might be blinking if it warned of pressing a key that would delete all data. Save the blinking effect for limited use on small graphics and always limit the number of times it blinks.

Font Selection

With all the font styles now available it can be tempting to use many of them in the same document; try to limit yourself to three font styles and three font sizes in a document. The idea of changing fonts is to add interest and to emphasize titles and subtitles. Three font styles and three sizes will allow you to achieve this without letting the screen become too confusing or cluttered looking. Confusing or cluttered screens can be more difficult to read and give users a negative impression of your documentation. There are dozens of fonts that work well on a screen, and dozens that do not.In addition, use standard fonts, as not all browsers support all fonts and must substitute something they can interpret.

Avoid fonts with:
- Thin or slanted strokes (as these distort on screens)
- Odd-shaped characters
- Crowded characters

Choose fonts that have:

- Long, easily-distinguishable lines on uppercase characters
- Round, well-proportioned single case characters
- Ordinary, easily-recognized characters

191

Chapter 9: Screen Design Considerations

Balance Fonts

When choosing which fonts to use, try to strike a balance between harmony and contrast, as this will add a professional look to your document. When choosing three fonts, choose:

- One font for the body of the text; avoid italics for this usage.
- One font that is an extended style of the body text font, such as a bold, narrow, or italic font.
- One font that is very different.

For example, for a business document you may choose:

- Times Roman, 12-point for the body
- **Times Roman, 12-point bold for subtitles**
- *Arial Narrow, 18-point italic for titles*

Your font choice may be influenced by specific requirements. For example, you may be designing Web pages for senior citizen vacation sites. You will want the fonts to be exciting, but easy for older eyes to read, so you may choose:

- Arial Rounded MT Bold, 12-point for the body
- Arial Rounded MT Bold, 18-point for titles
- **Modern, 14-point bold for subtitles**

Online Color Concepts

Introduction

Color preferences are highly personal and subjective, and vary according to one's species, culture, age, and sex. For example, birds, insects, and human babies are attracted to bright, garish colors; older humans prefer colors that are less bright. Therefore, there are no standards for what constitutes the perfect colors for online delivery. There are however, color guidelines that will help you achieve a polished, professional look that is easy to read and not too tiring on the eyes.

Color and Vision

The human eye brings different colors into focus at different distances behind the lens of the eye. As a result, some colors are brought into focus nearer the lens than other colors; these colors are called *cool* colors and have blue or blue tones. Colors that are brought into focus farther from the lens than cool colors are called *warm* colors and have red or yellow tones. As result of this phenomenon, warm colors appear closer than cool colors.

It is difficult for the human eye to focus on both warm and cool colors exactly at once, so when these colors are near each other, they give an illusion of depth. When warm and cool colors are next to each other, we often perceive them as being incongruous, or we say they "clash." Often this incongruous effect tires the eyes.

Cool Colors

Cool color include any that are based upon blue or green, such as: blue, purple, green, bluish-green, bluish-red, mauve, lavender, and magenta. There are cool tones of the warmer colors as well.

Warm Colors

Warm color tones include: red, orange, yellow, yellow-green, brown, and green. There are cool tones of warm colors as well. Humans tend to focus on warm colors more easily; babies are more attracted to warm colors, and even birds and insects are more attracted to plants that have warm colors. (Don't wear warm colors outdoors if you want to avoid bees and wasps!)

Neutral Colors

Neutral colors allow the most contrast with either cool or warm colors, and include black, gray, or white (although technically speaking, these are not colors). Beige and tan can be considered neutral. A neutral screen background can give your documentation the most clarity on a screen.

Limit the Numbers of Colors

For online reference systems, limit the number of colors used to three or four. Use these colors consistently throughout the system. For example, use the same colors for the same items on every screen. This will help users to memorize where to look and instill these colors as cues to finding those items quickly.

Color Contrast

For the best clarity of text on a screen, choose type and background colors that offer the most contrast. Avoid using type and background colors that have the same degree of lightness. For example, a yellow type on a white background is illegible, and navy blue type on a black background is very difficult to see. For very dark screens use yellow, bright green, or white type. For very light screens, use dark blue, black, or sometimes red type.

Colors to Use Carefully

Be careful in using pinks, pastels, and purples, unless you are sure the users' screens will reproduce them well. These colors tend to be distorted by some monitors. Use red sparingly as an accent color, as red can tire the eyes quickly.

Color and Culture

When choosing colors, be sensitive to cultural differences in what those colors symbolize; colors are symbolic messages that can subconsciously influence how your message is received and perceived. For example in

Chapter 9: Screen Design Considerations

Western society, white represents purity, goodness, cleanliness, youth, and unity (as in a white wedding dress). In some Asian societies, white represents old age, death, and mourning. If your online reference screens will be viewed in a culture other than your own, you may want to take the time to research what your color selections symbolize in that culture.

Color Impairment

Approximately 5% of the population is visually impaired when viewing colors. There are a few people who see no color whatsoever, but most people with color impairment have varying degrees of color impairment.

The most common impairment is the inability to distinguish varying shades of blues and greens. For this reason avoid contrasting two blues, or a blue and a green. Another common color impairment is the inability to distinguish between green and red. Choosing colors that contrast in their degrees of lightness and darkness will help readers who are color impaired to see contrasts better. For example, contrast light colors, such as tan, pink, or yellow, with dark colors, such as black, navy, or purple.

Color can be a very important tool to use in designing effective Web pages. Refer to the next chapter to understand online navigation.

Chapter 9: Screen Design Considerations

Chapter 10:
Elements of Online Navigation

Introduction

Now that you have a good understanding of basic online design, refer to this chapter to understand the elements of online navigation, which will make your online reference Web pages easy to use.

Online or Web documentation has a distinction that is unprecedented in the history of the written word: readers cannot judge the size of the online document by merely looking at it, as readers can do with a paper document. With a paper book, readers can lift it and judge its weight, flip through the pages to see if the book is filled with pictures or tiny print, and read information on the back of the book or cover flaps to see what the book is about. All of these options are absent with an online or corporate web document, so it is up to the developers to make the document easy to browse and navigate, and to give users a quick familiarity with the document.

Navigational Metaphors Definition

Navigation metaphors are online organizational designs that are likened to an organizational design in the real world. For example, an online manual's design may resemble a paper manual – and is even called a "manual" – to liken its organizational design to that of a book. Navigational metaphors are used when there are multiple volumes within the online or corporate web system from which to choose information.

Chapter 10: Elements of Online Navigation

The purpose of navigational metaphors is to allow users to relate the unseen computer design to something with which most people are familiar. Navigational metaphors give users familiarity and comfort with the online or corporate web document, and allow them to focus on content, rather than navigational structure. Most navigational metaphors are presented as graphic images that are linked to the subject's or volume's documentation. There are several common navigational metaphors in use:

- **Bookshelf:** in which volumes of the online system (or topics of one manual) are represented by a graphic displaying books sitting on a bookshelf
- **University:** in which different volumes or categories of information are represented by a graphic display of buildings that are labeled according to subject
- **Shopping mall:** in which different categories of information are represented by a graphic display of a shopping mall with "stores" labeled according to subject
- **Wheel:** in which different volumes, topics, or categories are arranged around the spoke of a wheel

These metaphors are used to organize large categories of information; often they contain sub volumes or subtopics accessed by clicking upon the graphic. For example, the university metaphor may have a science building. When users click on it, they see different science topic categories, such as physics and biology that may be located on "floors" of the building. When users click on a floor, they may see subtopics represented as rooms on the floor. Metaphors are not just pretty pictures upon a screen; each metaphor must be able to lead users to more detailed levels of information.

Point of Reference

Online reference systems do not automatically give users a point-of-reference from which they can determine where they are in the volume or topic. Web systems tend to resolve this problem more easily by having *back* buttons, *forward* buttons, and *home* buttons on the browsers, and relatively short topics in a Web frame.

Chapter 10: Elements of Online Navigation

A well-designed online screen has easily recognizable, intuitive point-of-reference clues that allow experienced users to navigate between screens within a couple of seconds without even thinking about it. A well-designed online screen should be easy to navigate for first-time users who have not had training. If you put the following information on each screen, you will greatly increase the likelihood of users' feeling comfortable with the online system:

- The volume, title, chapter, topic, and subtopics
- Important text
- Less-important text
- What are links and what are not links
- Where to go for more information

Standalone Information

Standalone Topics

In many online reference systems and in most Web systems, different topics of information are written as standalone topics. The word *standalone* is a technical term borrowed from computer networking; it means that the computer, or in this case the topic, can be used in isolation from other computers or topics.

Developing a standalone topic means to write the topic as a separate piece of information; users would not need to read other documents to understand the standalone topic in its entirety.

Benefits of Standalone Topics

Writing a standalone topic can be challenging, but this type of writing is rapidly becoming a new standard on the Internet and many corporate intranets and online reference systems. Writing standalone topics has many benefits, as standalone topics:

- Offer a complete, single subject of limited length that can be read quickly

- Can easily be maintained and rearranged within the system without impacting other topics
- Can easily be linked to other topics by use of hyperlinks
- Are easily browsed by users
- Offer hyperlinks to more detailed information or other topics

Challenges with Standalone Topics

There are also challenges with writing standalone topics, as they must be planned very carefully. The primary difficulties are in deciding:

- How large and how detailed each topic should be
- If there should be a limit on the length of each topic
- How to make each standalone topic relatively similar in length, detail, and consistency for a series of standalone topics
- If the topics are suitable for standalone presentation, or if each topic builds *knowledge* upon a previous topic

Tips for Writing Standalone Topics

The most important consideration for standalone topics is whether or not the type of information to be presented can naturally or easily be separated into individual topics of equal detail. Many types of information can be divided into separate topics and work well in standalone presentation. These types of information include:

- Procedures
- Processes
- Frequently Asked Questions (FAQs)
- Distinct sections of larger documents
- Topics that can satisfy one specific need
- Information on unrelated topics (such as a company's mission statement, financial reports, product information, customer service information)
- Catagorized information with easily recognizable distinctions (such as foods, countries, medical information, marketing information)

Many Web sites contain standalone topics in a menu on the left side of the screen. Each topic is hyperlinked to a Web page that discusses only that topic.

Information Not Presented as Standalone Topics

Some types of information do not work well as standalone topics. These types include:

- Information that is not easily separated into individual topics
- Information that must build upon a sequence of knowledge or information in order to be utilized
- Information that must be read sequentially for any reason
- Information that is very complicated and naturally connects to other topics

Chunking Information

Once you have decided to use standalone topics, it is important to have the information properly *chunked* in order to make distinct standalone topics. *Chunking* is an old technical-writing method whereby the information is subdivided into separate units that can usually be displayed on only one screen.

Label Each Chunk

The arrangement of topics in this book is an excellent example of chunking. Notice that the paragraph labels indicate one topic of thought, and that each topic is only several paragraphs long. Chunked information is referenced more quickly and read more easily if it is labeled; the label should express the main thought of the chunk or summarize its information. For example, this chunked paragraph is labeled *Label Each Chunk*; the label summarizes the idea of the paragraph.

Some writers prefer chunk sizes between one and seven paragraphs. Other writers prefer smaller, more succinct chunks that have one idea.

Use Chunk Labels to Verify Flow

Many writers scrutinize chunk labels to verify that the writing and its ideas have a nice flow. Each chunk should logically lead into the next chunk. If you analyze the flow of information by just reading the chunk labels, you will be able to see which chunks of information are out of place, or have a poor flow within the document.

Considerations for Chunking

A screen is a good indicator of a properly sized chunk, but screen size should never be the deciding factor in *what* to chunk and *how*. It is better to look at the information from the user's viewpoint and ask yourself:

- Where are the logical separators for this information?
- Can most of the information be chunked in a consistent manner (with similar length, similar detail of information, and a similar amount of complexity?)
- Where are the beginnings and endings for the topics' thoughts or concepts?
- Is there a logical start and finish to each topic?
- How would the user look up information?
- Would this chunk satisfy an initial inquiry?
- Should more details of the topic be hyperlinked to another chunk?

Tips for Chunking

There are many examples of chunking in this book: most of the chunks have distinct topics, with labels to indicate a change in thought. The following are other tips for chunking:

- Use one logical thought or concept for one paragraph.
- Do not use more than seven paragraphs per chunk of related information.
- Limit chunk size to one screen, or one to seven paragraphs. (If you think the chunk should be longer, you may really have *two* thoughts or concepts to chunk rather than one. Consider separating the two thoughts into two chunks.)

- If you are using cue cards, use a one-paragraph chunk per card.
- Use hyperlinks to connect chunks together.

Hyperlink Concepts

Hyperlink Definition

There are dozens of definitions for exactly what a *hyperlink* is. The easiest and broadest definition of a hyperlink is that it is an internal link within the system represented by and accessed by text or a graphic on the screen. When users double-click on a hyperlink it takes them to any of the following:

- Another place within the system
- Another place in a different system, such as a different Web site
- A smaller window with additional information overlaid upon the same screen
- The term *hyperlink* is a broad term that comprises two types of system links: *hypertext* and *hypergraph*.

Hypertext is simply text that is a link; upon clicking it with a mouse, the user is taken to another screen, or sometimes a drop-down box appears with more detailed information about the linked phrase or word. (Drop-down boxes often contain *definitions* of hypertexted phrases or words.) Many people use the terms *hyperlink* and *hypertext* interchangeably and use them to refer to all types of hyperlinks. A *hypergraph* is a graphic that serves as a hyperlink.

Underlining Hypertext

When writing in an online system or Web system, use underlining to denote hypertext. NEVER underline text for any other purpose; this is a globally accepted Web and online standard.

Underlining text for emphasis when it does not indicate a hyperlink will confuse and frustrate readers who expect the text to take them to another screen, and will give the impression that an amateur created the Web page or online page.

Hypergraph Definition

A hypergraph is a graphic, icon, or photo that initiates a hyperlink and takes the user to another screen. On web pages, hypergraphs are called *hotspots*. Because some graphics indicate hyperlinks and most do not, users look for changes in the mouse pointer to indicate a hypergraph. Usually the mouse pointer changes from a pointer to a hand with a pointing finger, although mouse pointers can vary widely and are designated by the mouse setup on the user's PC.

Hypermedia Definition

Graphics or icons usually represent hypermedia in an application and initiate hyperlinks that present information using motion or sound. For example, let us say that an online manual in a candy factory shows an icon to instruct factory workers in a specific step. When the icon is clicked, an animated cartoon appears that shows the physical movements and steps involved in pouring a ton of sugar into a huge vat. Hypermedia is being used more and more as the cost of storage space declines, and as software to create hypermedia is becoming less expensive and more sophisticated.

Other common uses for hypermedia include:

- Audio presentations of music or radio programs on the Internet
- Simulating classroom instructors in computer-based training courses
- Animating graphics on Web sites

Mouse Changes

When a user drags a mouse over hypertext or a hypergraph, the mouse pointer changes, often to a pointing hand or other pointer; this change indicates a hyperlink. The changing pointer is especially important for hypergraphs, as it is the only way users know if graphics indicate hyperlinks. Usually, the mouse pointer changes from a pointer to a hand with a pointing finger, although mouse pointers can vary widely and are designated by the mouse definitions on the individual PC.

Chapter 10: Elements of Online Navigation

Primary Advantage of Hyperlinks

In the ancient days of the very first online reference systems, hyperlinks did not exist. Online reference systems presented information in the same format and the same organization as paper manuals did. Users had to scroll with arrows keys (as PCs and mice did not exist) to get to different information.

Hyperlinks revolutionized online reference, as they allowed the online or Web design to be adapted to the information, rather than forcing information into a linear, sequential structure. The primary advantage of hyperlinks is that they allow the information to take on any organizational design, which allows writers enormous creativity. Hyperlinks allow a mixture of organizational structures to be used and linked, which allows the information type to shape the organization of information.

For example, an online listing of system error codes could be organized numerically and be hyperlinked to multiple topics organized in a corporate web structure. Both types of information could reside on different systems, and be written and maintained by different people who would need to coordinate changes in hyperlinks.

Other Advantages of Hyperlinks

Hyperlinks have many other advantages, including:

- Allowing different writers to work on different sections of a document and then linking the sections together easily
- Giving users more ways to access information
- Giving users more options of accessing more information
- Allowing information to be stored on different servers or systems and still appear to the user as seamlessly linked together

Primary Disadvantages of Hyperlinks

Hyperlinks can be a nuisance to maintain, especially hyperlinks between systems. Writers who maintain the hyperlinks must keep careful track of

Chapter 10: Elements of Online Navigation

the hyperlink source data and where the links take readers. One of the most common problems on Web sites is that the links are not maintained, and attempts to use the links result in an error. Because hyperlinks are often embedded in text, they can be inadvertently moved or deleted when text is changed. Therefore, it is critical to test all hyperlinks whenever changes are made to the system. This is especially critical if the hyperlink accesses another system.

In addition to maintenance, there are also other disadvantages of hyperlinks:

- If hypertext is used improperly when users expect *sequential* information, the flow of ideas can be disrupted and information can be difficult to understand.
- Users unfamiliar with online or Web reference, or users accessing a poorly designed system can get lost in the information and not know how to get back to the original information or their starting point.

These disadvantages can be remedied by careful design and by using hyperlinks appropriately.

Advice for Tracking Hyperlinks

Some writers find it helpful to keep written logs to track each hyperlink. Some online reference systems allow writers to display all hyperlink sources and even note which hyperlinks have changed. Most online reference systems at least allow a hyperlink log to be maintained, and a few of the better online reference systems automatically notify the system administrator when a hyperlink is not working properly. Some online reference systems and some Web design software automatically move the hyperlinks to the appropriate places when changes are made.

Hyperlink Most Information

Many, if not most other types of information work well when accessed by hyperlinks. These types of information include:

Chapter 10: Elements of Online Navigation

- Unrelated information; for example, a company's mission statement, contacts, product listing
- Related information that can be easily divided into sections, such as a table of contents where each chapter can be accessed by hyperlinks
- Sequential *lists* of information, such as catalog listings, or error codes
- Hierarchical information, in which lower levels of hyperlinks offer more detail on the same topic; for example, a department's organizational chart where managers are hyperlinked to individual subordinates
- Information in other formats (For example, if your document is mostly text, a spreadsheet could accessed by a hyperlink.)
- Tangential or anecdotal information that most readers would not be interested in, but which some readers would appreciate; for example, glossaries, historical information, case histories, financial figures

Naming Hyperlinks

Provide users pertinent information on what each hyperlink contains by naming them well. Giving hyperlinks funny or creative names may work well on the Internet when the goal of the hyperlink is to capture attention and lead the reader to an advertisement: even then, humor or creativity can confuse or offend readers from different cultures. Be funny or creative *only* if you are sure of the users' culture and when the hyperlink label can *still* give users accurate clues as to what the hyperlink contains. When naming hyperlinks, use simple terms that are understood by everyone.

Hyperlink Navigation

Introduction

This section discusses options for hyperlink navigation that you may want to consider. In hyperlink navigation an easy rule applies: The simpler the hyperlink design, the easier the hyperlinks are to maintain: the more sophisticated the hyperlink navigation is, the more difficult the hyperlinks are to maintain. Designing the right hyperlink navigation for your particular documentation is an excellent way to assure that users will not be lost and that the system will be easily maintained. Using hyperlink maps with any hyperlink navigational designs are another way of assuring that your users can navigate the system easily.

Hyperlink Maps

Most online reference systems and Web site design tools allow for hyperlink maps to be created. System administrators use these maps to maintain the hyperlinks, but the maps also help users keep track of where they are in the system. Maps are particularly helpful in large, complicated systems.

In most systems you can allocate a button from which users can access the map. These maps may take various forms, such as a standard organizational chart, a fishbone chart, or graphics of trees with different branches.

For example, you may want to show the map as a tree with labeled branches; the branch a user is on can be a different color. The map can be presented in a scroll box to take up less room on the screen, or you can allow users to link to the map. In this manner users would see a portion of the tree as they used arrows on a scroll box to lead them to the main trunk of the tree or to other branches. You can also use metaphors, such as universities and shopping malls as discussed earlier in this chapter. Hyperlink maps can be used with any of the following hyperlink navigational designs.

Sequential Hyperlinks

Sequential hyperlinks mean that the order of the hyperlinks is in a sequence, usually a logical or "natural" sequence. For example, a long document may have hyperlinks at the end of each chapter that lead to the next chapter. Usually sequential hyperlinks lead users in one direction only.

For example, if a document has five chapters, at the end of chapter one would be a hyperlink to chapter two, and at the end of chapter two would be a hyperlink to chapter three. The document would not have hyperlinks

Chapter 10: Elements of Online Navigation

between chapters one and three, or backward hyperlinks to a previous chapter.

The advantage of sequential hyperlinks is that they are very easy to design and maintain. They are also very adaptable to paper manuals that are being redesigned for online presentation with few organizational changes.

The disadvantage of sequential hyperlinks is that they do not give users options in which direction to go, or where to go next, as the hyperlinks lead in only one direction; therefore, they are not very useful for readers who want to *browse* information.

Determining exactly what previous knowledge should be read in sequence, or assuming what knowledge most users already know (and therefore does not need sequential presentation) can be a difficult judgment to make. In these cases, two options usually exist: force users to read the information sequentially, or supply hyperlinks and references to the earlier information for users who need it.

Hierarchical Hyperlinks

Hierarchical hyperlinks are the most common hyperlinks on Web sites and are often used in online reference systems. Hierarchical hyperlinks begin with broad, high-level topics that subdivide into lower-level topics. Usually the user can go backwards, at least back to the main menu. For example, on a corporate web site the hierarchical hyperlinks might look like this

Main Menu
  Corporate Policies
    Dress Code
    Ethics
  Human Resources
    Compensation and Benefits
      Annual Raises
      Vacation Policies

209

Chapter 10: Elements of Online Navigation

> **Payroll**
> **Tax Forms**
> **Payroll Correction Forms**

The advantage to hierarchical hyperlinks is that most users feel comfortable using them because of the inherent logic in hierarchies. The disadvantage in these hyperlinks is that they must be well designed and consider all possibilities. Hierarchical hyperlinks must anticipate what users will want and how they will think. Hierarchical hyperlinks are best used when there is only one logical way of organizing the hierarchy.

For most information, designing good hierarchical hyperlinks is not too difficult, although it can be confusing if there are multiple ways to organize the hierarchies. For example, programming languages could be organized under *Programming Languages* or under the system names for which they are used, such as *Mainframe Programming Languages,* and *Distributed Environment Programming Languages:* these could even be subhierarchies under *Programming Languages*. If there are too many "logical" ways to organize information, you may want to consider webbed hyperlinks.

Webbed Hyperlinks

Webbed hyperlinks are those with no *seemingly apparent pattern* to the hyperlinks; for example, it seems as if every hyperlink is linked to every other hyperlink, or that the hyperlinks have a random pattern. In reality, however, there is usually a pattern to the hyperlinks. Usually a webbed hyperlink system actually consists of a hierarchical system with many hyperlinks that cross reference other hierarchies, and/or sequential hyperlinks that cross reference other sequential or hierarchical hyperlinks.

The advantages to webbed hyperlinks are that they offer many navigational routes for users to take and that they are multidirectional. Webbed hyperlinking is the most sophisticated hyperlinking there is with current technology. A carefully designed webbed-hyperlinked system can make even the most complicated information very user-friendly. With webbed hyperlinking there are often no rules in what is linked to what; therefore, there is great creativity for writers in organizing and linking the information.

The disadvantage to webbed hyperlinking is that it is difficult to maintain. Great care must be taken to track the hyperlinks and to retest the validity of all hyperlinks when there are changes. Although tracking hyperlinks needs to be done regardless of what navigational design is used, webbed hyperlinks makes this task very cumbersome because of the cross-reference hyperlinks and the sheer number of hyperlinks this design uses.

Importance of Hyperlink Maintenance

All hyperlinks need to be tracked and maintained by some kind of process. The larger the system and the more complicated the hyperlinks, the more complicated the maintenance process will be. A disciplined approach to hyperlink maintenance is extremely important. Occasionally, in even the best-maintained systems, there will be a hyperlink that leads to nowhere. Repairing many hyperlinks is a very labor-intensive effort, and meanwhile users are being frustrated by broken links.

Remember: the primary reason users abandon Web sites is because the site's hyperlinks were poorly maintained.

A Hyperlink Maintenance Process

Therefore, it is advised that a person be assigned to maintain the hyperlinks, and that a maintenance process is written and followed. With the least sophisticated software and the smallest number of hyperlinks, the minimal process should include:

- A written record of where the hyperlinks are, their creation dates, the source of each hyperlink, and its target
- Manual testing of affected hyperlinks when changes are made
- Periodic testing of all hyperlinks in the system

Hyperlink Software Features

The process of maintaining hyperlinks can be simplified by using online reference software or Web site development software that has features for tracking and maintaining hyperlinks. The best software features include the ability to:

- Generate navigational maps automatically
- Generate file listings of hyperlink sources and hyperlink targets
- Notify you of any hyperlink changes and the location of each affected hyperlink

The features listed above are especially helpful when the system may be maintained by different people or will be very large. In the future, software will be able to move or edit the hyperlinks for you, but until that time, hyperlinks will need to be recreated manually when information is changed.

Many online reference systems and all web systems use markup languages to track hyperlinks. Markup languages are discussed in the next chapter.

Chapter 11:
Markup Language Concepts

Introduction

This chapter is intended for anyone who needs an understanding of what markup languages are, how they operate, and how they might add value to an organization. The use of markup languages is having a significant impact on technical writing and information technology.

Markup Language Definitions

1. A markup language is a set of codes added to text to define the structure of the text or the format in which it is to be displayed, stored, transmitted, or printed. The use of a markup language is a discipline to categorize and mark (or tag) different structures of information for the purpose of electronically organizing the structural components of the text.

2. A markup language can also be a component of a software program with specific program codes that mark, transmit, and store different structures of textual information. Markup codes can also be embedded in programs and data to enhance the ability of different systems to share data.

3. The term *markup* refers to the editorial or design marks that are handwritten onto a paper manuscript and that denote changes or instruction.

Chapter 11: Markup Language Concepts

Purpose of Markup Languages

During the early days of computers, all of the data that computers dealt with consisted of numbers (such as, 00011001110101010) and usually represented accounting information. Very little data represented text. As computers became smaller, their software became more sophisticated, and as PCs became more common, online reference systems and word processing systems proliferated. Textual information began to be compiled, stored, and transmitted as data to other computers. However, there were inherent difficulties with transmitting text-as-data to different computer systems with different machine programming languages.

Challenge with Text-as-Data

Text-as-data was recognized by computers as one long stream of words. The computers did not understand where titles, subtitles, paragraphs, font changes, print color changes, and other textual structures began and ended.

Therefore, when dealing with text-as-data there were two challenges:

- How to allow computers to *recognize* the various components and structures of textual information?
- How to *retain* the same components and structures of textual information when they were translated into other programming languages or transmitted electronically?

Markup languages were developed in response to these challenges, and markup language standards were also developed for using markup languages within textual structures. For one to understand the purpose of markup languages and their usage, it is helpful to have a clear understanding of textual structures.

Textual Structures

Textual structures are the building blocks of documents and are comprised of all the components that make up a document. There are many types of textual structures, but some examples are:

Chapter 11: Markup Language Concepts

- Titles, chapter headings, and subtitles
- Headers and footers
- Publication information
- Formatting information
- Author names
- Dates, page numbers, volume numbers
- Paragraphs and sentences
- Indexes and tables of content

Without a markup language, a computer would not understand where each of these items starts and ends, and it would not understand how to space these items or arrange them in the desired format when they are printed. In electronic printing, markup languages also tell computers what font to use, what colors to use, how to space these structures, and how to insert graphics.

The example on the next page is HTML code from a Web site. Notice how the heading is defined, how spacing is defined, and how much code is required before the actual text appears. This example was heavily edited and is intended to give you an idea of how a markup language looks. In reading this example, you can see how headers, titles, borders, formatting and spacing are defined.

Chapter 11: Markup Language Concepts

Example of a Markup Language

```
<HTML><HEAD>
<TITLE>WhiteFeatherPress.com </TITLE>
</HEAD>
<BODY bgcolor=#FFFFFF text=#000000 link=#000099 vlink=#990000>
<CENTER>
<table cellspacing=1 cellpadding=1 border=0 width=700>
<td align=right valign=top width=148>
<a href="/"><img src= "/img/head/logo.gif"
alt= "Feather" width=148>
<height=45 border=0></a>
<table cellspacing=1 cellpadding=1 border=0 width="100%">
<td valign=bottom width=552 height=31>
<font size=4 face="arial,geneva"><b> Books In Print</b></font></td>
<ALIGN="LEFT" VSPACE=13></A>
Welcome to WhiteFeatherPress.com, an innovative publishing company with its pulse on technical communication needs.
<BR CLEAR="All"><P>
```

Common Uses for Markup Languages

Originally, markup languages were used to help manage large amounts of text that proliferated in publishing companies, governments, and defense industries. As it became cost-effective to use online reference systems, and as online reference systems became more sophisticated, markup languages began to be used in other large industries (such as pharmaceutical, energy, aerospace, software) that depend on quick access to large amounts of information.

Markup languages are also the languages of object-oriented interfaces, database schemata, and transactions, and they allow unstructured data to

Chapter 11: Markup Language Concepts

be archived. However, markup languages used in these instances are beyond the scope of this book, which focuses only on how markup languages are used for managing online documentation.

Web Computing Uses Markup Languages

Over time, software and computers became increasingly more sophisticated, and web-based programs were developed for use in corporate web sites and Internet Web sites. Web pages use markup languages to control the structure and layout of the web page's text and graphics.

Markup Languages Can Reduce Access Costs

A primary benefit of corporate web-based programs is that they can often make large amounts of information accessible to many users without requiring a license for each user, and therefore offer low-cost access. With this advantage, companies have been able to move huge amounts of text to web-based online reference systems. Markup languages have facilitated the compilation, storage, display, transmission, and printing of that information. As web-based technology grows, so does the use of markup languages.

The use of the Internet has created an explosion of text transmitting online between different computers. Accompanying the explosive growth of the Internet, has been the huge growth of electronic commerce, which has furthered the use of markup languages to transmit text between businesses. Web-based programming, the Internet, and electronic commerce have all contributed to the exponential growth of markup language usage.

Document Management Systems and Markup Languages

Document management systems (DMSs) are another technology that is contributing to the use of markup languages. DMSs consist of complex software programs that compile, control, share, and manage a large number of documents. For example, a large insurance company with

217

Chapter 11: Markup Language Concepts

thousands of different contracts may use a DMS to manage and share different components of those contracts. A DMS would allow a sales representative to quickly extract different paragraphs of different contracts and compile them into a new contract customized for a specific customer. However, before different components of contracts can be extracted, their structure must be defined with a markup language.

Search Engines and Markup Languages

Search engines are web-based programs that search for the occurrence of specified words or phrases throughout thousands of documents. Search engines return to the user a list of where those specified words or phrases are in a list of documents.

Search engines are commonly used on the Internet and in corporate intranets as a primary way to search for information. They operate by searching for words within textual structures that are defined by markup languages. Without markup languages, search engines would not operate, and finding information on the Internet or private webs would be extremely difficult.

Advantages of Using a Markup Language

There are many advantages to using a markup language, as it:

- Allows highly-defined components of text to be shared, controlled, transmitted, and managed.
- Allows textual structures to be transmitted between different computer systems while retaining their structures, formats, and colors.
- Can allow different writers to reuse text, or to work on the same document simultaneously.
- Allows search engines to operate.
- Saves readers enormous amounts of time searching for information within a system managed by a markup language.
- Provides quick access to information and thereby enhances corporate knowledge, which in turns leads to faster production and a more informed workforce.

Chapter 11: Markup Language Concepts

- Lends a method to access and control information for a large amount of documentation, thereby making the information more usable, meaningful, and valuable.
- Allows companies to control and disseminate their business information quickly, thereby increasing the value of that information as a corporate asset.
- Lends companies a competitive advantage by facilitating access to information, which allows management to make quicker and more informed business decisions.

Disadvantages of Using Markup Languages

There are also disadvantages to using a markup language, as it requires:

- Funding, training, resources, and time to implement.
- Corporate commitment to its implementation, maintenance, and governance of its use.
- Discipline, standards, processes, and procedures, to implement and maintain the markup language.
- Cultural change to accept the markup language as the preferred method of enhancing the use of documents.

Common Markup Languages

The long-term benefits that markup languages offer to an organization are enormous, and the use of markup languages is proliferating. The development of markup languages has had an interesting history and has been driven by innovations in technology and changes in the world marketplace. The first markup language created was SGML.

SGML

Many industries that sell products internationally must adhere to rigorous international standards. Early in the 1980s the International Standards Organization (ISO) began using the Standard Generalized Markup Language (SGML). SGML was invented by Dr. Charles F. Goldfarb with

Chapter 11: Markup Language Concepts

the IBM Almaden Research Center, and was later known as ISO 8879. This standard became widely used in governments and industries, as it allowed computers with different machine languages and coding schemes to communicate textual structures in an internationally agreed upon way. SGML is still one of the finest and most detailed markup languages available and may be a preferred choice if your organization:

- Desires International Standards Organization (ISO) certification.
- Interfaces with governments.
- Interfaces with the following industries: defense, energy, financial, scientific, or aerospace.
- Does business internationally.

Outside of these types of organizations, however, SGML never gained the popularity of some newer markup languages, namely, HTML and XML. This is because:

- There were comparatively few software programs developed to exploit the benefits of SGML.
- Due to the rigor with SGML must be deployed, it was never widely used on the Internet: the Internet itselfs tends to popularize whatever technology is used on it.

SGML had its greatest popularity before the Internet and is best suited for very large and highly-disciplined organizations. The organizations largely responsible for the explosive growth of the Internet in the early 1990s were small and creative. These organizations chose easier, less disciplined markup languages, especially HTML.

World Wide Web Consortium

The World Wide Web Consortium (W3C) started in 1994 with a primary goal of developing protocols that facilitate consistency in the development of Web sites. The consortium develops guidelines, specifications, software and tools to facilitate Internet comunication, including guidance on many markup languages. The tools are especially helpful, as they can assist

Chapter 11: Markup Language Concepts

you with editing or validating your code. Before your organization choses a markup language, or if you are new to HTML, XHTML, XML, or other commonly used mark up langages, you may want to explore their wonderful Web site at http://www.w3.org.

HTML

Hypertext Markup Language (HTML) gained very wide popularity in the early years of the Internet, and most Internet web pages are still created using HTML. HTML is comparatively easy to learn and use. It has few standards and was therefore favored by early developers of Internet Web pages. (An excellent Web site that offers HTML utility tools is www.tucows.com.) HTML is superb for easy tracking and marking of hypertext. HTML is very user-friendly, but it cannot communicate with different computer applications and databases the way that XML can.

XHTML

XHTML is the first major change to HTML since version 4.0. It is based upon HTML coding and format, but it adds markup tags that allow the HTML code to be read by all browsers, including those used in hand-held computers, wireless devices, televisions, and automobile computers. XHTML reformulates HTML as an XML application. You can use a utility called HTML Tidy to convert your HTML documents to XHTML. HTML Tidy is an open source software (meaning it can operate on different platforms) and it is free as long as you comply with the copyright notice and license. You can download HTML Tidy from http://www.w3.org.

XML

Extensible Markup Language (XML) is based upon SGML. XML is (and will continue to be) one of the most important technologies for sharing information in the early years of this new century. XML is the first meta-data markup language to be supported by other technologies.

Chapter 11: Markup Language Concepts

Meta-data means information about data, and refers to the information that defines how data is arranged, stored, and accessed. It includes information such as column names, row names, or numbers, or whether the data is a flat file or an object. This meta-data is needed for a system to retrieve specific information.

XML is also the first meta-linguistic markup language. *Meta-linguistic* refers to information about the data's language, and it defines the language's syntax. Information about data language syntax allows XML to access and retrieve information from any unknown data language, as long as the language has an XML interface.

Because of its unusual traits, XML has a tremendous impact on information management and documentation management. XML allows users to retrieve information without knowing how it is organized or what data language it is in. XML allows different computer systems with different languages and different operating systems to share information, as long as those systems have an XML front-end interface, which acts as connectors between XML and the data.

Different computer systems are being connected by the use of XML, and to date, no other markup language can compete with it in this way. For these reasons, XML may be the markup language of choice for your organization, and it is being widely used on the Internet and on intranets. XML is also being integrated into new software by major software vendors and database vendors. For those readers who are interested in learning more about XML, contact the Organization for the Advancement of Structured Information Standards (OASIS); they have a portal to information on XML's impact on documentation and application development, as well as information on other markup languages. At this writing their Web site is located at http://www.oasis-open.org/.

Wireless Markup Language
Wireless Markup Language (WML) is a specialized markup language created to allow electronic commerce on wireless personal devices, such

Chapter 11: Markup Language Concepts

as mobile phones, personal data access devices, and hand-held computers. WML is based upon HTML and uses an encoder to translate the information into WML. WML was developed to meet the challenges of high-data latency and low-transmission speeds of wireless technology, and accompanies the use of Wireless Application Protocol (WAP) used for applications that execute on wireless personal devices.

WAP and WML are widely used in European countries for mobile banking, but their use has not yet been widespread in the United States of America or Canada. As mobile bank and wireless access to the Internet gain popularity in North America, the implementation of WAP and WML will be very widespread. If you expect your information to be displayed on wireless devices you may want to consider XHTML, XML, WAP, and WML for your wireless applications and documentation.

Choosing a Markup Language

If your organization has decided to use a markup language, there are many things to consider. Even though most of these considerations are beyond the scope of this book, here are a few tips. Before choosing a specific markup language for your organization, do a great deal of research. Talk to different vendors who offer markup language interfaces with their products. Understand the capabilities of each markup language as you consider it, and understand how the markup language will need to be maintained.

Define the documentation management requirements of your organization. Be sure all of your requirements are well defined: be especially sure that your organization has defined the future directions of information management and data sharing, and that you have management's commitment to the implementation.

Chapter 11: Markup Language Concepts

Markup Language Return-on-Investment

The use of a markup language, an online reference, or a corporate web system can greatly improve the flow of information through an organization, help to provide more accurate information, and improve organizational efficiency. When implemented correctly, the return on investment for these technologies is typically 12-18 months, well within typical expectations for a good (if not excellent) investment.

Using a Markup Language is a Discipline

The use of a markup language, an online reference, or corporate web system is as much a discipline as it is a technology. These systems must be maintained in order to be an asset to the organization: they cannot be installed and ignored. Without maintenance, these systems will degrade in quality, and with that degradation go the confidence of the users in the system's usefulness and the decline of its use.

People tend to forget that information is transient; it changes with events, with process changes, and with priorities. People also tend to ignore the fact that information imbedded in computer systems is nothing more than magnetic code whose ionic charges make it adhere to plastic; it can easily be erased by static, electrical charges, human error, and all sorts of system failures.

Maintenance Responsibilities

All these are compelling reasons to maintain the online reference or web system. Experience has shown that online systems are maintained better and more cheaply when one department has funding, skilled people, and authority to:

- Create and enforce information standards, design standards, and writing standards.
- Create and communicate backup and maintenance processes.
- Determine when and how changes to the information will be updated

Chapter 11: Markup Language Concepts

- Serve as the single point of contact with the technologists who support the computer and network systems.
- Administer author rights to people in the company who want to submit information into the system.
- Edit people's submissions.
- Determine when and how new versions of the software will be acquired and installed.

With management's support, funding, and the right skills, maintaining an online reference or web system is not difficult. It requires discipline, common sense, and a basic understanding of computers and their software. Most importantly, it is the maintenance process that determines how long the information in a system will meet the needs of an organization and whether or not the system will be perceived as a long-term success.

Chapter 11: Markup Language Concepts

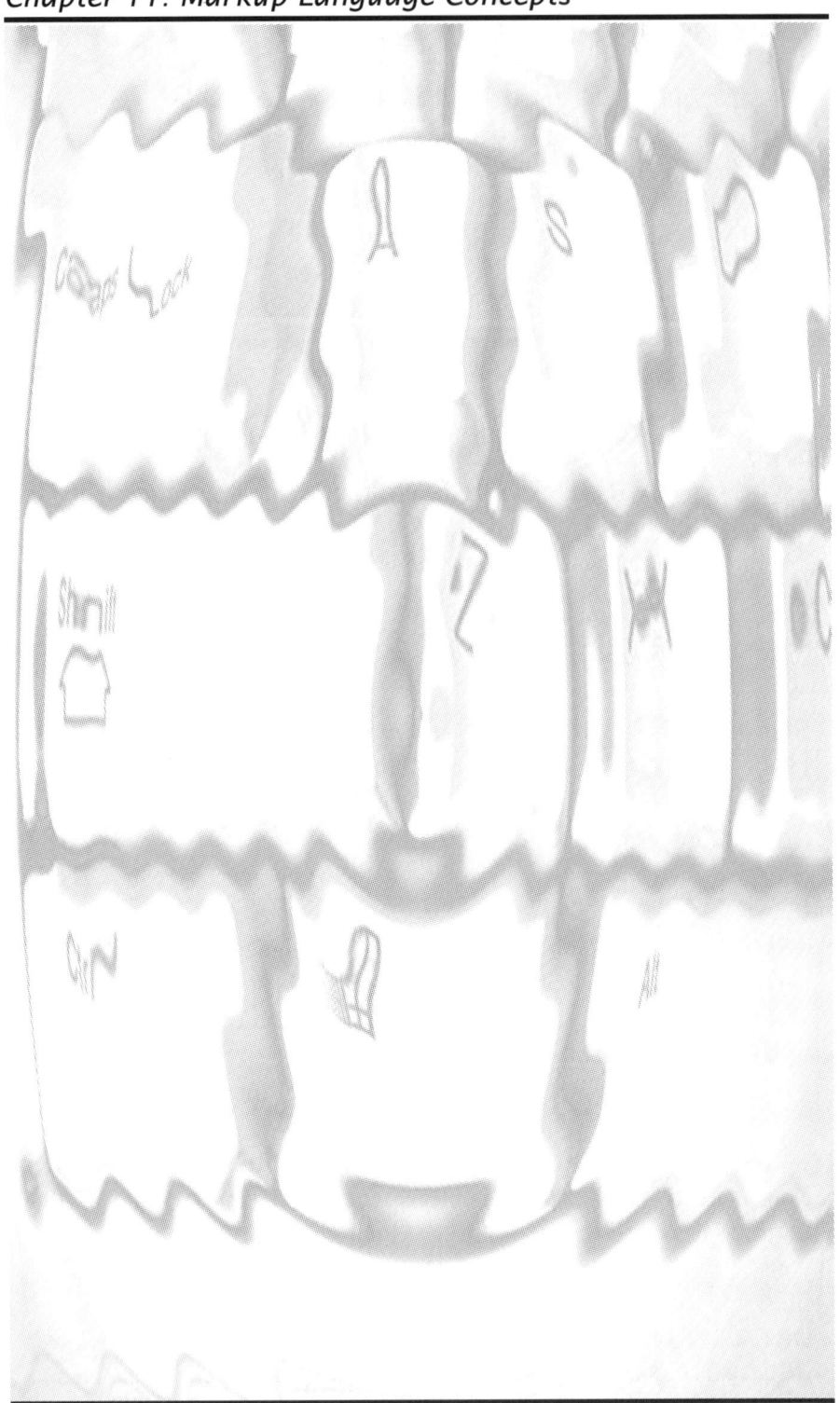

Appendices

Project Time Estimation Procedure

Introduction

This procedure is for anyone who needs to estimate how long a technical writing project will take. It should yield an estimate within 30% or better of the actual project time. Use paper and a calculator when following this procedure.

Step 1: Estimate the Interview and Meeting Time

Make a list of everyone from whom you will need to gather information. Include programmers, business users, team members, trainers, and managers. Plan an initial one-hour meeting with each contact, add up the time, and label this sum A on the paper.

After you know how long the entire project will be, come back to this list and add in one half to one hour more per person, per week for either meetings or phone calls. This time also takes into account the time spent waiting for returned phone calls and rescheduled meetings. Also add in vacation times and several days for illness or emergencies.

Step 2: Estimate the Reading or Research Time

Make a list of your paper and online resources. They may include the Internet, other in-house documents, manuals, or reference books. If your references are in softcopy, you may need to merge hundreds of files, and then edit or update them, so give this step careful thought. Divide this resource list into those you need to read thoroughly, and

Appendices

those you only need to browse. You will probably choose to read them in increments. Most people are good at estimating the time it takes them to read material. Estimate the time you think you will need to browse the material, and then the time it will take to read the material thoroughly.

Add up your estimated reading time. Add in another third of the reading time to allow for interruptions. Label this sum B, and write it under sum A. If you have no resources and will do all of your own research, include at least three full days for this. Be generous in your estimation. Label this sum B, and write it under sum A.

Step 3: Estimate the Time for User Analysis
Add in two days for conducting a user analysis if all users will be local. Add in four days for conducting a user analysis if users are local and remote. Label this sum C and write it under sum B. (Refer to the section entitled *User Analysis* in *Chapter 1* for more information.) Delete this step if no user analysis will be conducted, and proceed to step four.

Step 4: Estimate the Time to Learn Software
Include this step if you are learning new software, such as new online help or reference software. Refer to the software manual, a trainer, or others who have used the software to estimate how long it will take you to learn the basics. Don't expect to master the software; you'll have this opportunity in the project work itself. Most people experienced with teaching themselves software can learn the basics of an online system in a day or two. Add up this time, label it sum D and write it under sum C.

Step 5: Estimate the Document's Size
Use this step to estimate the length of the document. Analyze your resources and consider the scope of the project. For example, will the documentation have 50-70 long procedures in a paper manual, or 10-15 short procedures on cue cards? Will the documentation be an in depth analysis of the technical architecture for new contract developers, or a software proposal for executives? (Tips for analyzing and organizing information are in *Chapter 2*.)

Does your organization have a style guide to follow? If not, do you want to take notes to use in developing a style guide for people to use later?

Carefully consider the project's scope, and outline to estimate the length of the document. Consider whether or not you'll have many graphics or pictures in the documentation, and whether you'll only be importing them or having to create them yourself.

For each freelance graphic you create, estimate 30-60 minutes per graphic, depending upon its complexity and your familiarity with the software. Try to find graphics that you can use instead of creating them yourself.

Rule of Thumb for Document Size

Are there documents similar to your project available for you to browse? How long are they? Use this guide to estimate the length of your document. Paper delivery assumes one inch margins.

For paper delivery:

- One paragraph of about 100 words without graphics will be about 11 KB in size.
- One page of text without graphics will have about 900 words, or about 13 KB
- Ten pages of text without graphics will have about 5300 words, or about 45 KB.

For online delivery:

- One page designed for online delivery with generous white space (like this book) but without many graphics will have about 400 words, or 15 KB.
- Ten pages of online text with generous white space and few graphics will have about 3000 words or 35 KB.

Notice that simple computations of text multiplied by the number of words or pages don't give you these results. This is because most of the kilobytes

Appendices

(KB) used in software is for empty space and formatting. Therefore, a full page of text is only slightly larger than a paragraph of text in KB.

Also notice that these measurements don't include graphics. How much software space or KB each graphic takes is dependent upon the graphic's electronic format and its size. Write down the estimated length of the document off to one side of the column of sums; label it L. You will use this figure at the end of the next step.

Step 6: Estimate the Writing Time
Use this step to estimate the actual time it will take to write the documentation.

Paper: If your L figure represents pages of information that aren't too technical, estimate 65 minutes per page. If the information is very technical, estimate 90 minutes per page. If the information is politically sensitive but not technical, estimate 60 minutes per page.

Cue cards: If your L figure represents cue cards, and if you know the software you'll be using to write them, estimate 5 minutes per cue card. If you are learning the software as you write, plan 10 minutes each for the first 25 cue cards and 7 minutes for each cue card after that.

Online: No rule of thumb exists for estimating the amount of time it takes to write online, as the online software and screen design influences how much information is on each screen. Most software allows the design of a "master template" that automatically designs each screen for you. The use of a mark-up language also influences the amount of time used in writing online documentation.

Generally, three to four online screens have about the same amount of information as one page, and take 45, 60, or 90 minutes to write depending upon the complexity of the content. Multiply the physical length of the document (item L calculated in step 5) by the estimated amount of time for writing that type of document. You'll probably want to convert this number to hours or business days. Label this sum E and write it under sum D. If you are a less-experienced writer, add another 25-35% of this time to sum E.

Step 7: Estimate the Writing Time
Add up sums A through E. Convert the time to days. See Step 7 in the sample project-planning sheet that follows this procedure. Estimate a six-hour working day to allow for the usual work interruptions. Label this "Writing Time."

Step 8: Estimate the Planning Time
How much time should you expect to spend planning a project?

Rule of Thumb:
- **If approaching a subject you know nothing about**, as many contract technical writers do, expect to spend a third of your project time planning the project and gathering information.

- **If the project calls for a stringent level of technical detail,** expect to spend half of the project time planning. Lengthen this time even more if you are new to the organization, or working in a fairly chaotic environment.

- **If approaching a subject with which you feel comfortable, and that doesn't require a stringent level of technical detail,** expect to plan roughly one quarter of your time planning and gathering information. Use the above guidelines to decide how much time that you will use planning.

Divide the "Writing Time" number by a factor of two, three, or four, add this quotient to the "Writing Time," and cross out the old "Writing Time" number.

Step 9: Estimating the Total Project Time
Now that you have the total number of weeks spent on the project, return to Step 1. Multiply the number of contacts in Step 1 with another

Appendices

half-hour or hour per week. Add this time to "Writing Time." This will give you an estimate of the total project time.

After using this procedure you will have a good estimation of how long the project will take: at worst you'll probably be about 30% off target: at best, 5%. Your outline and scope may shorten, shift, or grow. Make certain your manager understands this calculation is an estimate. With experience, your estimations will become more accurate. You may find it helpful to create a Project Time Estimation Worksheet, as shown on the next page.

Appendices

Example of a Project Estimation Worksheet

Project Estimation Worksheet		July 1, 2002
	Melissa Manning and Jamal Lewis	PH: 12345
	This project will create a UNIX development guide to provide naming standards, library conventions and messaging information for the UNIX platform. Local UNIX developers will use the guide.	
Step 1	Time spent in interviews and meetings, with 8 contacts: time not consecutive	Sum A = 4 days
Step 2	Time spent reading sources	Sum B = 4 days
Step 3	Time spent for local user analysis	Sum C = 0
Step 4	Time spent learning XYZ software	Sum D = 1 day
Step 5	Estimated length of document, L = 100 pages	Length = 100 pages
Step 6	Estimated length of writing time multiplied by pages: 60 minutes x 100 pages	Sum E = 6000 minutes
Step 7	Estimated writing time: 6000 minutes divided by 60 minutes per hour = 100 hours: 100 divided by 6 hour days = 17 days	Sum F = 17 days
Step 8	Estimated planning time for technical content.	Sum G = 6 days
Step 9	Add up the days in progress thus far. Add additional meeting time for Step 1. Add 3 days additional time for illness, vacation, and emergencies. Divide days by a 5-day week.	32 days 2 days +3 days ------------ 37 days
	Total project time	**7 weeks, 2 days**
ECD:	**August 25, 2002**	

233

Appendices

Common Acronyms List

In the field of IT there are many acronyms, and the most common are listed below. Those items noted with "don't write out" shouldn't be written out when first presented, unless you think your readers will have a special need to know the words in the acronym. Note the use of capital letters or lower case letters.

ANSI
American National Standards Institute (Don't write out.)

ASCII
American Standard Code for Information Exchange (Don't write out.)

API
Application programming interface

ATM
Asynchronous transfer mode

BBS
Bulletin board system

BIFF
Binary interchange file format

BIOS
Basic input/output system (Don't write out.)

CASE
Computer-Aided Systems Engineering

CBT
Computer-based training

CD
Compact disc (Don't write out.)

CD-ROM
Compact disc read-only memory (Don't write out.)

CD-RW
Compact disc read and write

COM
Component object model

CPU
Central processing unit (Don't write out.)

CRT
Cathode-ray (Don't write out.)

CSR
Customer service representative

DBMS
Database management system

DCOM
Distributed component object model

DDE
Dynamic data exchange

DIF
Data interchange format

DLL
Dynamic-link library (.dll is accepted in file name.)

Appendices

DOS
Disk operating system (Don't write out.)

EISA
Extended Industry Standard Architecture (Don't write out.)

FAT
File allocation table

Fax or **fax**
Facsimile (Don't write out and don't use all capitals, as in *FAX*.)

FTP
File Transfer Protocol (Use ftp for Internet location.)

GPI
Graphics programming interface

GUI
Graphical user interface

GUID
Globally unique identifier

HAL
Hardware abstraction layer

HBA
Host bus adapter

HMA
High memory area

HTML
Hypertext Markup Language

HTTP
Hypertext Transfer Protocol (Use "http:" in Internet addresses.)

IBN
Interactive broadband network

IEEE
Institute of Electrical and Electronics Engineers, Inc.

I/O
Input/output (Don't write out.)

IPS
Integrated Performance Support

IS
Information systems

ISA
Industry Standard Architecture (Don't write out.)

ISP
Internet service provider

ISV
Independent software vendor

ITV
Interactive television

KCOM
Kernel component object model

LAN
Local area network

Appendices

LCD
Liquid crystal display

MAPI
Messaging application programming interface

MCA
Micro Channel Architecture (IBM Trademark)

MIDI
Musical instrument digital interface (Don't write out.)

MIPS
Millions of instructions per second (Don't write out.)

MIS
Management information systems

NAN
Not a number

NDIS
Network driver interface specification

Net BEUI
Network basic input/output system extended user interface

Net BIOS
Network basic input/output system

OCR
Optical character recognition

ODBC
Open database connectivity

ODL
Object Description Language

OO
Object oriented

OLAP
Online Analytical Processing

PIF
Program Information File

PSU
Power supply unit

QA
Quality assurance

RAID
Retrieval and information database

RAM
Random access memory

RDBMS
Relational database management system

ROM
Read-only memory

RTF
Rich Text Format

SCSI
Small computer system interface

Appendices

SDLC
Synchronous data link control

SGML
Standard Generalized Markup Language

SIC
Standard industry classification

SIG
Special interest group

SNA
Systems network architecture

SNMP
Simple network management protocol

SPI
Service provider interface

SQL
Structured Query Language

SVGA
Super video graphics adapter (Write "Super VGA.")

TCP/IP
Transport Control Protocol/Internet Protocol (Don't write out.)

TIFF
Tagged image file format

TSR
Terminate-and-stay-resident

UMB
Upper memory block

UNC
Universal naming convention

UPC
Universal product code

UPS
Uninterruptible power supply

URL
Uniform Resource Locator (Don't write out.)

VAR
Value-added reseller

VGA
Video graphics adapter

VM
Virtual memory

VSAM
Virtual storage access method

WAN
Wide area network

XHTML
Extensible Hypertext Markup Language

XML
Extensible Markup Language

Appendices

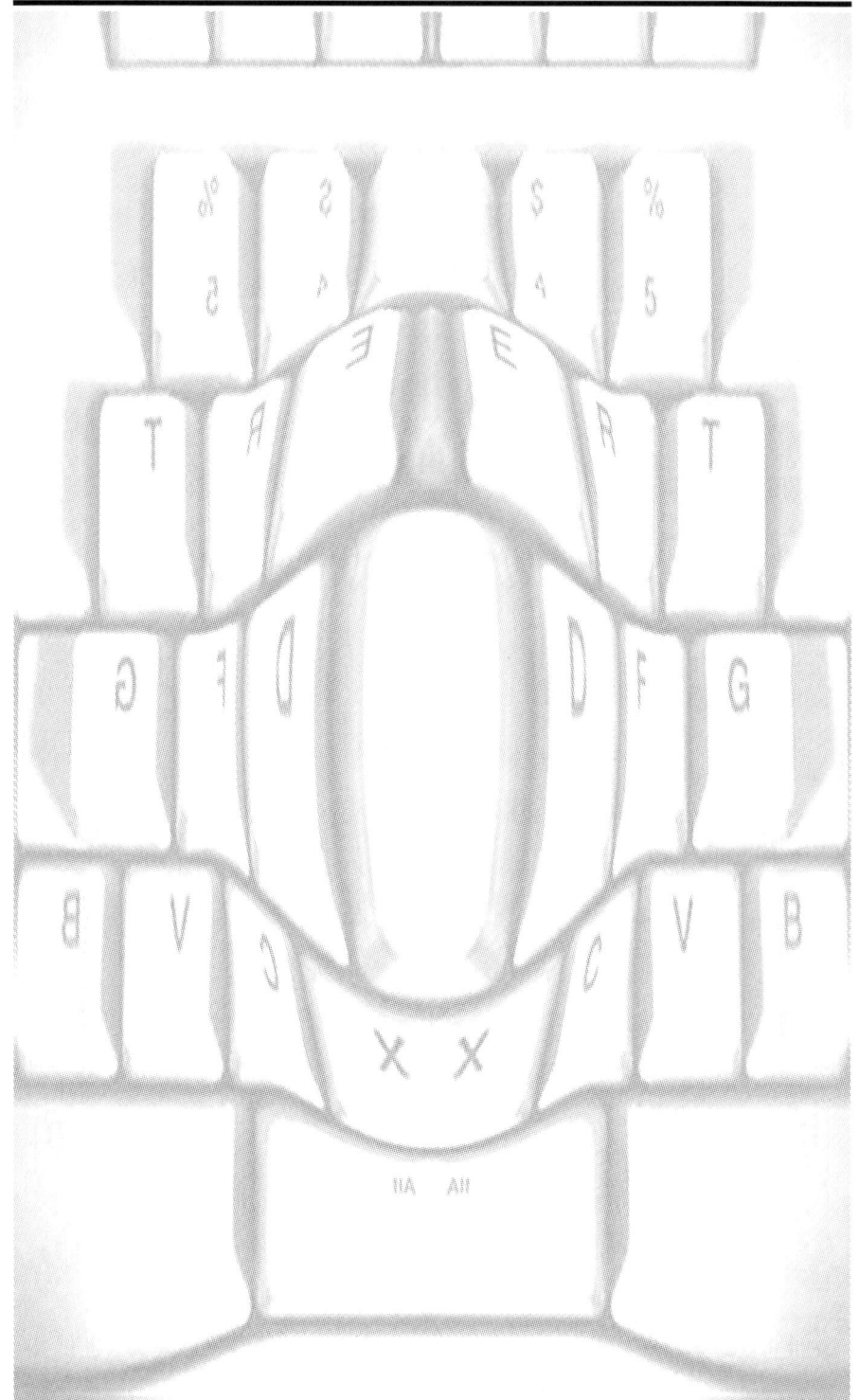

Index

A
acronyms
    examples of 89
    how to use 87
    plurals in 89
    rules for 88
    using online 89
active and passive person
    examples 84
active person
    definition 84
active vs. passive 83
appendix
    definition of 47
audience
    considerations for defining 26
audience intention statement 21

B
background
    definition of 45
bibliographies
    purpose of 120
bibliography
    example of 121
    how to punctuate 120

Index

Bits per second 102
blinking font 191
bold fonts
    on screens 189
bottom margin area 183
bottom margin frame 183
bps 102
browse list
    example of 39
    purpose of 38
bulleted lists
    examples of 97
    using 95
bullets
    using 95

C

capital letters
    guidelines for screens 190
capitalization
    and numbers 112
    in indexes 135
capitalized letters on screen 189
capitalizing
    deities 111
    geographic locations 110
    nationalities 110
    people's titles 109
    rules for 109
    special events 110
    words in titles 111
CBT 159
choosing colors 194
choosing fonts 192
choosing system structure 167
chronological hierarchy 51

Index

chunking
    considerations for 202
    definition 201
citing multiple references 119
Colons
    with quotations 106
colons
    in sentences 105
    in titles 106
    rules for using 105
    with hours/minutes 105
    with ratios 106
color concepts 192
color impairment 195
Commas
    using 103
commas
    in a sequence 103
    with clauses 103
    with numbers 104
compound number 100
computer memory
    abbreviations for 102
computer-based training 159
concordance indexing
    use of 127
content frame 184
content listing area 184
conventions
    for writing indexes 132
cue card
    examples of 170
    procedure example 172
Cue cards
    definition 170
cue cards 157

Index

    and procedures 172
    and system structures 171
    types of information 158
    uses for 157

D

dashes
    use of 115
decimals 101
designing
    for both paper and online 181
designing online screens 187
Development Phase
    of planning 4
distinct-categories hierarchy 52
DMS 218
document management systems
    and indexing 127
    and markup languages 217
document size estimation 228
double negatives 92

E

electronic books 145
elements of online navigation 197
ellipses
    how to use 107
endnotes
    examples of 118
    use of 117
    vs. footnotes 117
estimating document size 228
estimating interview time 227
estimating project time 227
estimating writing time 230
executive summaries

definition of 47
executive summary
　　determining length of 48
expert users
　　expectations of 25
Extensible Markup Language 221

F

fishbone structure 163
font
　　bolding on screens 189
　　italics on screens 189
　　online guidelines 188
　　sizes on paper 148
font blinking 191
font colors 190
font selection 191
footnote numbers 118
footnotes
　　examples of 118
　　use of 118
　　vs. endnotes 117
foreign language 86
foreign phrases
　　avoiding 86
　　when to use 87
formula
　　to determine links 166
frames
　　types of 182

G

GB 102
gender inclusive language 85
genders
　　avoiding he or she 85

Index

Gigabit 102
Gigabyte 102
grid structure 164

H

help authoring tools 158
help systems 158
heterogeneous audience 20
hierarchical hyperlinks 209
hierarchies
    of organization 50
hierarchy
    chronological 51
    distinct-categories 52
    least-to-complex 50
    most-to-least complex 51
    sequential 52
hierarchy structure 162
hints
    for clear writing 94
homogeneous audience 20
HTML 221
Hungarian Notation 89
hypergraph
    definition 204
hyperlink
    definitions 203
hyperlink maps 208
hyperlink navigation 207
hyperlinks 209
    advantages of 205
    and underlining 203
    disadvantages 206
    hierarchical 209
    maintenance 211
    naming 207

Index

    sequential 208
    webbed 210
hypermedia
    definition 204
hypertext
    definition 203
Hypertext Markup Language 221
hyphen
    use of 112
hyphenating numerals 101
hyphens
    and compound modifiers 113
    and compound phrases 113
    and compound words 112
    and prefixes 114

I

if/then procedure example 173
implementation phase
    of planning 5
indentions
    with indexes 132
index
    audience considerations 141
    considerations for content 140
    determining what to 138
    format 132
    how to 138
    length 131
    regeneration 131
index capitalization 135
index columns 132
index cross-reference 134
index font 134
index page numbers 133
index pointers

Index

definition 130
index punctuation 136
index styles 133
indexes
    indentions 132
    other considerations for 141
    purpose of 123
indexing
    and online reference systems 124
    approaches 139
    choosing words to index 143
    concordance 127
    estimating time 142
    impact on system resources 124
    limited-word 129
    types impact systems 126
    use of limited-word 129
indexing jargon 137
indexing manually 143
information
    human processing 41
information type
    deciding 41
Internet references
    in endnotes, footnotes 119
introduction
    example of 46
it, it's
    use of 92
italics
    on screens 189

J

jargon
    avoiding 90
    definition of 90

    types of 90
    usage 91
just-in-time presentation
    definition 42

K

K 102
Kbps 102
kHz 102
Kilobit 102
Kilobits per second 102
Kilobyte 102
kilobyte 102
Kilohertz 102
KWIC
    impact on writers 128

L

limited-word indexing
    impact on writers 129
    use of 129
links
    determining number of 166

M

maintaining links 165
markup language
    advantages of 218
    choosing a 223
    definition 213
    disadvantage of 219
    example of 215
    purpose 214
    return on investment 224
    using a 224
markup languages

Index

common uses for 216
MB 102
Megabit 102
Megabyte 102
Megahertz 102
mHz 102
mouse changes 204

N

navigation metaphors
 definition 197
navigational metaphors
 purpose of 197
neutral colors 194
novice users
 expectations of 23
numbered lists
 using 99
numbers
 compound 100
 how to write 100
 hyphens in 101
 multiple 101
 ranges of 101
 using letters for 102
 when to spell out 100

O

OASIS 222
online delivery
 benefits of 155
online font guidelines 188
online reference
 definition 156
 design considerations 179
online reference index system

Index

    managing an 130
online reference index systems
    importance of maintenance 129
online reference systems
    types of 156
online source documents
    considerations for 37
organizational
    culture 15
outline
    how to write 58
    purpose of 55
overviews
    definition of 45

P

page
    numbering on screens 181
page format 147
paper
    advantages of 146
    considerations for delivery 147
    disadvantages of 146
    impact of 145
paragraph labels 151
passive person
    definition 83
person
    first 81
    second 81
    third 82
persons
    use of 81
planning
    considerations for 3
    development phase 4

Index

    other considerations for 8
    risks of not planning 1
    verification phase 5
point-of-reference clues 199
possession
    with 's 91
possessive pronouns 91
prefixes
    and hyphens 114
primary area 183
primary frame 183
procedure
    definition 43
    if/then example 173
procedures
    and cue cards 172
    and system structures 174
    titles for 43
procedures vs. processes 175
process
    definition 44, 175
    example 175
    example of 44
processes
    and readers 176
    and system structures 177
project plans
    purpose of 2

Q

quotation marks
    use of 116
quotations
    within quotations 116

R

reading environment 22

S

s
   when used in possession 91
sans serif fonts
   and online viewing 188
scope document
   content of 11
   purpose of 9
screen
   numbering 181
   resolution 181
screen design considerations 187
screen layout 180
screen navigation 180
screen resolution
   and fonts 188
screen shape 180
screen space 180
screen viewing area
   definition 182
screen viewing areas
   types of 182
screens
   multiple screen considerations 181
   vs. paper 179
search engines
   and markup languages 218
second person
   definition 81
   example of 81
secondary area 184
secondary frame 184
semicolon
   used in indexing 136
semicolons
   in lists 108

Index

    in salutations 108
    in sentences 107
    purpose of 107
sentence
    hints for writing 75
sequential hierarchy 52
sequential hyperlinks 208
sequential structure 161
Serif
    definition 148
SGML 219
side labels
    use of 151
Sign-offs 30
source documentation
    gathering 33
source documents
    considerations for 36
    online 37
standalone
    definition 199
standalone topics 199
Standard Generalized Markup Language 219
structure
    choosing a structure 167
structures
    multiple 166
sub-index
    indention 133
Subject matter experts 28
system pointer problem
    example 130
system pointers
    definition 130
system structure
    definition 160

Index

    purpose of 160
system structure maintenance 168
system structures
    and procedures 174
    and processes 177
    types of 160

T

table of contents
    impact on index 140
TB 103
Terabit 103
Terabyte 103
text-as-data 214
textual structures
    definition 214
third area 185
third frame 185
third person
    definition 82
    example of 82
titles
    for reports 69
    for tutorials 70
    for user guides 70
    hints for writing 71
    options for procedures 70
    writing effective 68
top margin area 183
top margin frame 183
topic labels 151
transfer users
    expectations of 25

U

underlining fonts 190

Index

underlining hyperlinks 203
underlining letters 190
uppercase fonts on screen 189
user analyses
    techniques for 17
users' point-of-reference 198

V

Verification Phase
    of planning 5
version control
    planning for 6

W

WAP 223
warm colors 193
Web
    when to capitalize 120
web structure 165
webbed hyperlinks 210
white space
    definition 149
    guidelines for using 150
    use of 149
Wireless Application Protocol 223
Wireless Markup Language 222
WML 223
writing styles
    purpose for 73
writing time 231

X

XML 221

Order Form

Price: $25 USD $29 CN   Ŀ26 UK   € 29 Euros

Order Online: Visit us at *www.whitefeatherpress.com* to order or to purchase the electronic PDF version for use with *Adobe Acrobat Reader*™.

Postal Orders: Complete this form and mail it with your payment to:

> WhiteFeatherPress.com., Inc.
> Attn. Book Form Orders
> P.O. Box 23549
> Jacksonville, FL 32241-3549

Or fill out the credit card information and fax this page to 1-904-260-3089

Please send me ___ copies of *Information Technology's Writing Survival Guide.*

SHIP TO:

Name _____

Address _____

_____

_____

FL Sales Tax: Please add 6.0% sales tax for books *shipped to Florida addresses.*

Shipping: Please add $4 shipping and handling charges for the first book, and $2 for each additional book to the same address. Select payment method below:

___ *Personal Check* ___ *Visa* ___ *Mastercard:* **Expiration Date:** _____

Name on credit card _____
___ Same
Billing
Address _____
___ Same

_____

_ _ _ _ _ _ _ _ _ _ _

Signature (Credit card orders will not be processed without your signature.)

Thank you!

```
Price ___ x ___ copies = _____ .
S & H ___ x ___ copies = _____ .
FL sales tax @ 6.0%    = _____ .

Total                  = _____ .
```